MW01697536

Not All Okies Are White

Not All Okies Are White

The Lives of Black Cotton Pickers in Arizona

Geta LeSeur

University of Missouri Press
Columbia and London

Copyright © 2000 by
The Curators of the University of Missouri
University of Missouri Press, Columbia, Missouri 65201
Printed and bound in the United States of America
All rights reserved
5 4 3 2 1   04 03 02 01 00

Library of Congress Cataloging-in-Publication Data

LeSeur, Geta J.
Not all Okies are white : the lives of Black cotton pickers in Arizona / Geta LeSeur.
p. cm.
Includes bibliographical references and index.
ISBN 0-8262-1271-9 (alk. paper)
1. Cotton farmers—Arizona—Interviews. 2. Cotton farmers—Arizona—Biography. 3. Afro-American agricultural laborers—Arizona—Interviews. 4. Afro-American agricultural laborers—Arizona—Biography. I. Title.
HD8039.C662 U65 2000
331.6'9960730791'09041—dc21   99-057338

∞™ This paper meets the requirements of the
American National Standard for Permanence of Paper
for Printed Library Materials, Z39.48, 1984.

Designer: Stephanie Foley
Typesetter: BookComp, Inc.
Printer and binder: Thomson-Shore, Inc.
Typefaces: Franklin Gothic and Minion

for those who have worked “in the cotton”—
who gave so much and got so little.

with respect and love

Midway

I've come this far to freedom and I won't turn back.
I'm climbing to the highway from my old dirt track.
  I'm coming and I'm going
  And I'm stretching and I'm growing
And I'll reap what I've been sowing or my skin's not black.

I've prayed and slaved and waited and I've sung my song.
You've bled me and you've starved me but I've still grown strong.
  You've lashed me and you've treed me
  And you've everything but freed me,
But in time you'll know you need me and it won't be long.

I've seen the daylight breaking high above the bough.
I've found my destination and I've made my vow,
  So whether you abhor me
  Or deride me or ignore me,
Mighty mountains loom before me and I won't stop now.

Naomi Long Madgett

Contents

IV. "We Are Sojourner's Daughters"

V. "Turn Not Away from Their Smiles and Their Tears"

A Note about Rememberin', Tellin', and Writin' by Lynn C. Lewis

This book by Geta LeSeur is the first to present the oral narratives of several members of a small Arizona community named Randolph that formed in the decades between 1930 and 1960. Through these individual narratives, a clear picture is painted of the realities of life for the working people who were the essential component of agricultural production as this section of the western United States developed, expanded, and contracted with technological change.

There is history in these narratives, a history inadequately documented in traditional textbooks. In one sense, this book is an extension of the study of the experiences of southern blacks as workers in the production of the staple crops that generated so much wealth for the American nation and the white landowners, and so little compensation to the workers themselves. Focusing on particular families and individuals in one community allows this collection to provide vivid personal stories that resonate with multigenerational presences and influences. As these narrators recount their work experiences, an important witness is finally given voice—that of the workers in cotton fields and tobacco fields; of those harvesting corn, rice, fruit, vegetables, with feet or knees touching the soil, doing the hand-to-stalk labor that brought all this bounty to fruition. Seen in this way, these narratives reflect a relationship to land that is akin to the old southern pattern of life tied to location and defined by geographic place.

In another sense, this book connects with another kind of experience, the relocation of human groups migrating to find subsistence and substance. In this sense, these accounts afford a new glimpse into the western expansion of the United States in the first half of the twentieth century. Migrant communities developed creative strategies for adapting to new realities and for interacting with diverse types of people and demands. The contracting of their labor is one of many interesting means utilized by black people to coordinate their efforts and to negotiate economic transactions with landowners.

Yet, in another sense, these oral narratives and LeSeur's contextualizing essays comprise an innovative blending of history telling and literary analysis. The narratives are oral histories, and the essays and headnotes are introductory ethnographies that offer a window on the time and place being documented. As such, *Not All Okies Are White* gives evidence of the resilience and adaptability of these migrant people and communities. People of African and European ancestry, Mexican Americans, and Native Americans from many groups, primarily the Pima, Papago, and Navajo, worked in close proximity yet lived—in the era before integration—separate lives, except for the occasional intermarrying that creates truly multicultural families. Although irrevocable losses, such as the loss of Native American languages and other traditions, occurred along the way, these narratives preserve the personal and collective memory of all of these groups, as exemplified in the lives of the eighteen individuals speaking here.

Additionally, there is a nineteenth voice in this communal text, one entrusted by the Arizona narrators to carry and interpret their stories. This voice is as interesting and memorable as the others because of its perspective on and passion for this subject. LeSeur writes about her subject with tenderness, and it becomes clear why these valley folks felt they could share their life histories with her. Because she questioned so many and listened so much, we should be compelled to listen to this text, to listen closely. Listen for the intricate interactions among groups; listen to the sense of humor; hear what hunger can cause an infant to do; notice how economic and technological forces impact women and families, while affecting differently the African, Native/Indian, and Mexican people.

Listen to the history in this text. If this book were a song, I believe you would enjoy both the words and the music.

Acknowledgments

I owe much gratitude to many different people. The *Arizona Daily Star,* which in 1980 did a story on Randolph, Arizona, made me aware of the existence of a black town in the state and a special population of migrants who, like the white Okies, ended up as pickers of cotton and other crops. Earl Epps, a former administrator in the Coolidge school system with whom I worked on a black studies project in the Tucson School District, shared many stories, anecdotes, and key information about black life in Coolidge and Randolph. I thank him for the family and individual names he provided to begin the personal contacts for the interviews that came later.

Katie Montaño, a federal agent at the McFarlane State Park in Florence, also suggested persons in the Casa Grande Valley to speak with as well as sources to consult. I also thank my husband, Ed, for finding Ms. Montaño for me in the early stages. Additionally, he meticulously sought out other persons, dates, and places on maps throughout the research. I thank the University of Missouri for a 1991 Development Grant that enabled me to begin my many trips to Arizona and to search for and purchase archival materials. A 1992 an American Council of Learned Societies Grant-in-Aid provided funding for research and visits to the National Archives to locate materials and persons knowledgeable about the cotton industry.

A 1994 Arizona Humanities Council Study Grant helped tremendously in the process of amassing the interviews and taping and transcribing the words of the individuals whose voices are heard in this book. I am grateful also for the council's inclusion of my project among the talks offered via its Speakers Bureau. That opportunity enabled me to present a work in progress and educate a new audience about black migrant life in Arizona. The Missouri Humanities Council Speakers Bureau also provided similar opportunities to share my materials. I thank it for that support, too.

The Arizona Historical Society in Tucson, the Casa Grande Valley Historical Society, and the Florence and Coolidge historical societies were all gracious and eager to assist me in locating materials and directed me to exhibits and files throughout the researching and compiling of this project. I especially want to thank Craig Ringer, director of the Casa Grande Valley Historical Society, for allowing me to borrow and use pictures, clippings, and transcriptions and for unlocking the museum for me to visit the society even when it was officially closed. Craig also invited me to publish the piece on Myrtle Jordan in *Casa Grande Valley Histories* (1996). Archivist Kay Benedict is thanked for the transcripts of former teachers, which were useful for the chapter on migrant education.

I am grateful to the *Arizona Republic, Arizona Daily Star, Tucson Citizen, Phoenix Gazette, Coolidge Examiner,* and *Casa Grande Valley Dispatch* for free use of pictures and articles. The National Archives granted me permission to use prints from its collection of Dorothea Lange's vivid Depression-era photographs of Arizona migrant life.

Special thanks also go to those who encouraged me to continue this project when, at times, it seemed futile, unappreciated, and unending. They include Arizona journalist Abe Chanin, whose *This Land, These Voices* chronicles the experiences of Arizonans from a multicultural perspective, including Jews; Bruce Dinges, director of the Arizona Historical Society Press; Carolyn Dorsey and Darlene Clark Hine, who saw the study of black women's position in the American West as critical for an understanding of black history in its entirety; and University of Missouri student Obie Blackmon, from Southern Missouri "Bootheel" sharecropping roots, who told me many stories of his family and of being poor, black, and a "cropper," while doing a capstone research paper with me on the southern Missouri strikes of 1939. There are also many others—colleagues, students, friends, and family—who always asked how the book was coming and smiled, always, at its title.

Of course I cannot thank enough four women who have worked very hard, under many time constraints, on transcribing and meticulously keyboarding the text over a long period of time. They are Janice Earle, Marilyn Keefer, Marilynn Keil, and Lynn Lewis. All were drawn into the stories and lives as they worked. I must, however, single out Lynn C. Lewis, a former graduate student in oral traditions at the University of Missouri and currently an assistant professor of English at Tennessee State University, for her encouragement, support, and hard work throughout. She assisted me not only as an editor but also as an adviser who provided insights about voice, diction, language, and technical details. A true professional, she at times made me sleepless because of her desire for perfection. But I shall always remember her assistance and involvement with this project with esteem and warmth.

Truth be told, after all is said and done, my biggest debt is to the people of Randolph, Eloy, Coolidge, and the entire Casa Grande Valley who allowed me into their lives and homes. A chair, a soda, a glass of water, a picture, a clipping, an obituary, some news item or potential lead was always forthcoming. The Jordans, especially Myrtle (now deceased) and Ron, my informant, resident historian, and guide; the Reeces; the Smiths; the Moores; the Fieldses; and the Whites—all have given me much, materially and spiritually, over the years. I consider them family. To them, and to the generations who follow them, I give my sincerest gratitude and love.

Not All Okies Are White

Prologue

I have always had an interest in oppressed peoples—the hardworking people, the people who build and plant and harvest: the underclass, the underfed, the underhoused, the underpaid, the undereducated, the despised, and the displaced. I came to this project on black migrant workers, therefore, with considerable emotion, drive, and empathy for these special folks, who number in the millions.

I come from a background of hardworking West Indian great-great-grandparents, grandparents, parents, and extended family. From that small village of Mocho, Clarendon Parish, Jamaica, West Indies, I, in retrospect, learned to recognize the nobility and pride work brings. I saw and remember how proud my grandmother was, dressed in her starched and ironed headkerchief, print dress, clean apron, stockings, and brown oxford shoes, as she placed her basket of fruits and vegetables on her head to go to market on Fridays. My grandfather would stand quietly by, watching her as she strode down the road. She would sell everything and use the money to buy pretty fabrics for sewing, tobacco for her pipe, household items, and "sweeties" for us—her grandchildren and our cousins.

Work defines who we are, and work is important in our lives. My aunt, who is ninety-five years old and lives in rural Jamaica, plans her life and works her land as if she were a thirty-year-old woman; my mother, who is eighty years old, works four days a week, by choice, and refers to some women younger than she as "little old ladies"! Work, I believe, gives us hope and keeps us moving forward; not working leaves us with despair and hopelessness. I have been working since the age of fourteen, because that is what I saw everyone in my family doing, and it has made all the difference in my notion of how life can and should be lived. I hasten to add, however, that the value of labor depends on the type of work and the circumstances. Forced slave labor, child labor, and poor conditions with little or no pay are not what I am referring to here, and the narratives herein will indeed illustrate some inhumane experiences with work. I truly believe that black people are an earthbound people, compatible with the land until costly modern technology, the breakup of families, an unequal economy, and opportunities denied affected them negatively.

Early into this project, about 1990, some of my colleagues questioned why I, as one immersed in English literary studies, would be doing oral history. They cited, often, my lack of credentials to do so. My response to them, most of the time, was that people of African descent cannot wait for or trust others to always recover our history for us; we have to find a way to "walk the walk" and "do the deed" when no one else has or will. I would also hasten to add that this often means that we have to teach ourselves and learn during the process how best to resurrect and mine our own buried histories. I've never pretended to be an anthropologist or sociologist, by academic definitions. Perhaps what I am is a "literary medium," because my major aim in this work is to capture and record the stories of Arizona's black migrants through the oral gifts they possess, and pass them on to an audience unaware of the multiplicity of their migrant experience.

A generation ignorant about African American history needs to know that twentieth-century life for blacks goes beyond the Mississippi River and the Mason-Dixon Line, that in some Western states African Americans experienced the isolations and oppressions of the slave era in places such as Kansas, Texas, Arkansas, Oklahoma, and Arizona, not that long ago. They also need to know that African Americans have maintained a vast amount of their culture in "cowboy territory." They need to know that American history is not complete without the inclusion of the black exodus to the American West. They need to know that, often, women and the elders were the backbones of these places and that in many ways their lives have not changed. They should realize that black communities have had continuity—culturally, religiously, and economically—in small, isolated towns like those in the Casa Grande Valley of Arizona. I call these folks black Okies, because not all Okies are white!

I came to this project rather naively and fearfully. I had read a story about Randolph, Arizona, in Tucson's *Arizona Daily Star* in 1980,[1] and I was amazed at the existence of a sizable "black town" that was more than fifty years old at the time. That photo essay showed the African Americans' involvement with the local cotton industry, which is one of the largest in the world. These black Okies had come there as pickers, beginning in the 1930s. My other amazement was that cotton actually grew in Arizona. I had always thought of cotton as being found primarily in the southern United States and in South and Central

Courtesy Arizona Daily Star

America and Africa. I simply had to find this place called Randolph, to see if it really had blacks and to learn more about it. I didn't go there right away, but my curiosity remained. Finally, in the late 1980s, on a hot summer day, I made the drive north from Tucson on Interstate 10.

Randolph, located in Pinal County, is south of Phoenix, east of Interstate 10, west of Coolidge, and sits among places with such exotic names as Tiara Palms Estate and Desert Beach and such descriptive names as Eleven Mile Corner Road. It is eleven miles from that corner to the towns of Casa Grande, Coolidge, and Florence, the county seat. Randolph has a population of approximately three hundred African Americans who came primarily from Kansas, Oklahoma, Texas, and Arkansas, and most of them belong to America's economic underclass. It was a strange scene that I discovered two miles from La Palma, a "small spot" in the road on Highway 87, about one-half mile from the Arizona Training Center. Randolph was stark, flat, quiet, brown, with empty lots, old cars, modest colorless-looking frame houses, some trailers, no mailboxes, no school, and a few paved streets. There was a strange and eerie quiet like that of a ghost town. Maybe it was the torrid summer heat that kept the residents inside; maybe no one was home. As I drove through, slowly, I discovered a small convenience store and stopped; it was indeed operating. This was not another Arizona ghost town.

Everyone in Randolph knows when a stranger enters the town or an "outside" car comes through. You may not see them, but they see you. They pay attention to everything, especially the children, who go from street to street to play or visit. Randolph, to me, was like a village, not unlike the one I emigrated from in the Caribbean, where everyone takes care of everyone, and almost everyone is related by blood, marriage, kinship, or place. As I struck up

Mandy Selmon Moore and her store, which closed in 1995.

a conversation with Mandy Moore, the shopkeeper and store owner, I began to understand better the nature of this town. From her store window, Mandy can see just about everything happening in and around the Randolph homes and streets. I felt that she was also important as the one who grants clearance to proceed with anything you are going to do in Randolph. Her store was the "check-in" place, and I never failed to stop there, chat for a while, buy a soda, and catch up on the town's news each time I visited. I was sorry to see it close in 1995; with it, a piece of Randolph died too.

I have always had a strong interest in black women in the American West and did a summer project in the late 1970s on black women pioneers in Arizona. Not much had been written about the African American contribution to the Western economy, either. There are various essays and books on the black cowboys and the Buffalo soldiers, but no one has chronicled their laboring activities and the role of black women in that process. What struck me about Randolph was the strong and evident presence of black women in the town, and since I am a professor of Black Studies and Women Studies at the University of Missouri, it was incumbent on me to really sink into this aspect of Randolph life. At that point, I decided to interview women only, so as to reconstruct their migrant experience, and did so in an essay published in 1989 for the Black History Month issue of *Compendium*.[2] I used no names, but simply drew on my impression of the women and the things I had heard and learned standing in Mandy's shop. More detailed work would require getting to know the residents better in order to gain their trust, confidence, and permission.

Because I taught at the University of Missouri and traveled to Arizona only during holiday breaks and summers, it was difficult to establish any continuity or build close relationships, and at times I felt like an intruder in their dust. Although I thought this was important work, I felt some guilt about entering into their already burdened lives, especially when someone didn't feel well or had something else to do, which was often. However, in spite of these encumbrances, over time I developed an excellent bond with several families. These relationships grew to the point that they would worry when they did not see me for a few weeks or months. I was no longer "that lady" to the grandchildren who often announced my arrival, but Geta or "the teacher." That always brought joy to my heart. Their kinship lines had now attached me to them.

I began by interviewing several older women whom I thought represented a cross section of the Randolph experience. One was Fannie Reece, from Hugo, Oklahoma, whose picture I had seen in the *Arizona Daily Star*. I was drawn to Mrs. Reece quickly, because she reminded me of my grandmother Arabella Ashley; her coloring, height, elegant head, prominent cheekbones, and twinkling eyes were like my Grandma's. Mrs. Reece had picked cotton all her life under difficult family circumstances. Then there was Mamie Selmon from Eufaula, Oklahoma, who never picked cotton but came to join family members. She is the mother of Mandy, the storekeeper. Others interviewed were Metris White, who came in 1936 with her widowed father from Wellington, Texas; Myrtle Jordan, a Pima Indian woman from the nearby Gila River Reservation who married a black man, Cleveland "Preacher" Jordan, from Hope, Arkansas; Edna Fields, from Marlin, Texas, the wife of Obed C. Fields, also from Texas, who came to buy land and farm, not to pick anybody's cotton but their own. And Sarah Lucille Smith, the white postmaster, who had been in Randolph for more than forty years. Mrs. Smith, although white, was an important link to the black community because she knew everybody's "business." She was the recorder of births and deaths and kept track of and documented all kinds of personal matters. She had a strong woman's bond to the black women of

Randolph, including the midwives and others who had those hard women's times. She also gave me many leads and provided a good, objective profile of Randolph and the surrounding communities. In fact, her narrative is the basis for most of the history of early Randolph included in this book. She was a very bright and capable woman who proudly announced that her name had appeared in *Who's Who of American Women.* Thinking all along that Randolph was a 100 percent black town, I wondered why she was still there.

I was curious about everything but quite timid about asking questions. I did not use a tape recorder initially, thinking that most feared "the machine"; I took notes instead. Not only was I missing important information while doing this, but I heard the skepticism in their voices as they tried to see what I was writing. Finally, I began using a tape recorder and reassured them that it was my way of getting everything right. The best taped interviews were those that were handled like conversations. Not only did this approach make everyone more comfortable, but it also allowed me to listen more keenly while keeping eye contact. The primary questions asked were: Why and how did you come to Randolph? What was the journey like? What was life like for you and your family? What was life like working with cotton? Why have you stayed? What made you leave your home state or town? How did you adapt culturally, regionally, racially? What was your relationship with Hispanics, whites, and Indians? What were the educational opportunities for you and your children? What was Randolph like then and now? What is most memorable for you about the various stages of your life? Of course, many other questions were asked and much information gleaned without solicitation as trust evolved over time. Often those conversations were held under unusual circumstances, and they took place in a variety of settings—in a church, in a kitchen, at a bedside, under a tree, or riding in a truck. It was never boring. Often grandchildren and sons and daughters and various other relatives would pass through during the interviews, so mixed voices can be heard in the background of my tapes. Sometimes the young ones would gather and listen as the elders related their experiences. They would interrupt the narrative to ask questions or add something, and I never discouraged them—these were their family histories.

Even though I made a special effort to interview only the elderly women, I found that the perspectives of some of the second-generation children and of the men were equally colorful and important. In the case of the family of Myrtle Jordan, Ron, the eldest son, remembered much about growing up black, Indian, and light skinned, which made life tough for him racially, culturally, and educationally. The passion and sincerity of his story bring tears to my eyes even now. I also taped his younger brother, Jeff, to get varying perspectives. Jeff's activism for change in Randolph was triggered when the county permitted the building of factories that dump toxic wastes in and around the community. Nick White, from Okmulgee, Oklahoma, is the pastor of the town's Church of God in Christ, and his memory for detail is amazing. His story and that of his wife, Metris, complement and enrich each other. Although I used the questions outlined earlier to guide me, they were not all used in every instance because uncharted information would come forth with each narrative. There were no boundaries and I had everything to learn—the mosaic was multifaceted. Often an interview with one person would lead me to seek someone else. Because these people did not always live in Randolph, I found myself driving to the surrounding communities of Coolidge, Casa Grande, Florence, Eloy, Chandler, Phoenix, Mobile, Cashion, or Sacaton on the Pima Indian Reservation.

Over and over, the overwhelming emphasis in the interviews was on the trip West from all those places they had lived before. The trips were often described as adventuresome, and these were their most vivid memories. I imagine that since these "travelers" were young when the journey West took place, they felt, like Metris White or the Joad children in John Steinbeck's *The Grapes of Wrath,* that it was fun to get on the road and ride to a new place, imagining good things and the romance of it all. Metris speaks, for example, of Indians in long dresses, animals by the campfire at night, and winding mountain roads that titillated her and her siblings' minds. Some told a different, more "believable" story of leaving in the dead of night with everything they owned because they had "some difficulties" with the white boss in Oklahoma or Texas. One of the other things they would mention, uniformly, was their hard work with cotton, boasting often of being a good and fast picker, bagging between five hundred and a thousand pounds per day and making good money. They would say, "Oooh honey, I could sure pick!" They also spoke of the days when Randolph was a flourishing town with stores, businesses, and a night life, when blacks "had it good"; a time when families were closer. They mentioned local heroes, such as Beulah Moore, a single mother who had a small contracting business and made enough money to buy property and send her kids to college; L. B. Reece, who fought for the welterweight championship at Madison Square Garden; Melvin Moore, who became a deputy sheriff in Coolidge; or a girl who was a track star in her high school; or Negro teachers who really cared about the kids, teachers like Matthew and Elgie Batteau, who taught at the Borree Corner and Eleven Mile Corner Road Schools, when segregation and the need to work prevented blacks from attending white schools.

Overwhelmingly, everyone mentioned two events that changed their lives forever in Randolph. One was the advent of running water and the water tank that was dedicated in 1963; the other, the arrival of automated cotton picking machines in the late 1950s. The machines did not discriminate; all workers of all colors and races felt its bite. They say those machines were responsible for the decline of the town, of their lives, and of everything else. The differences in the narratives were evident only when individuals spoke about their particular family circumstances. All had come to Randolph with the same goal: to work, earn a living, buy a place, raise a family as best they could, and settle, or move on.

One misconception I had early on was that Randolph was totally black—it's not now, but it was predominantly black in the 1940s and 1950s. To get a balanced picture of the community, I also interviewed the Acunas—"Los Acunas"—a Mexican family who had been in Randolph for more than forty years and who had raised their children and picked, chopped, and irrigated the cotton fields alongside Indians and blacks.

In Randolph, one notices not just economic segregation but also educational and geographic segregation. Highway 87 runs north from Interstate 10 to Coolidge and divides Randolph into east and west—colored and white. It was zoned that way in the 1920s. On the west side of the highway are the few remaining whites—the Smiths, the Woods, and others. One young white girl once told me, "I live on the white side of Randolph." On the northeast and southwest are a smattering of Mexicans like the Acunas, and to the east and at the center of Randolph proper are the blacks. They, of course, outnumber the other racial groups. Although the numbers fluctuate depending on who leaves, returns, dies, or sets up a trailer overnight, they may be broken down thus: 250 blacks, 20 Mexicans, 30 whites.[3]

Courtesy of the Arizona State Historical Society Library

Epes Randolph, for whom the town of Randolph is named.

The whites have been there the longest, since the early 1920s, when Colonel Epes Randolph, for whom the town is named, envisioned a sprawling, successful city not far from Casa Grande. The other racial groups arrived periodically as the "old ones" died or property was bought or repossessed. There are no signs of disharmony in this village, though. People seem to get along. Their assimilation from the South, Midwest, Southwest, or Mexico could perhaps serve as a model for American acculturation.

In spite of the geographic boundaries, all the houses and yards have a similarity about them. It would be difficult to say which are owned by whites, by Mexican Americans, or by blacks. There are also some marriages across racial lines, like that of the Jordans, which perhaps resulted naturally from long years of working, playing, and going to school together. On the colored side of town the community bonds are strong and traditional, with family and gender roles clearly defined. Mexican and black bonds are also close because of work, school, and proximity. Perhaps the better term for Randolph now is "community of color."

Although this "story" began by attempting to document the uniqueness of Randolph as a black migrant town whose history was not to be found in libraries, museums, newspapers, or other archives, the residents' voices became the source of almost all the information I gathered. So, the town's story and their life stories became one. Other sources in the surrounding communities—such as Eloy, where I found Parlee Bradley—extended and enriched the Randolph story. Mrs. Bradley and her husband, James, were featured in a *National Geographic* essay, "Cotton, King of Fibers," in the June 1994 issue. In it she says, "Everywhere you looked was cotton, cotton, cotton. . . . we could clean a stalk in just a few seconds."[4] I decided that this was a woman I had to meet. Her story was colorful, and, coincidentally, she had worked for the Jordans of Randolph when they were contracting for the white farmers.

The important topics of migrant education and the African American experience with cotton and its contribution to the economy of western Arizona led me to Matthew and Elgie Batteau of Tucson. They taught the Randolph children in the segregated schools at Borree Corner and Eleven Mile Corner Road during the height of cotton farming of the late 1940s and early 1950s. Thousands of migrant children received inadequate education during those years not only because of segregation but also because of the itinerant nature of their lives. Many dropped out for good, and many remained illiterate, which obviously hindered their ability to find better employment opportunities when cotton "died."

This study historicizes the interesting, unique, unknown, and misunderstood things about Randolph, Arizona, and selected communities in the Casa

Grande Valley from the 1930s to the 1960s. It also describes what life with cotton was like for an African American community that was essentially invisible to the region, state, and country. It attempts to find cultural linkages and heritage lines in order to document the quality of life for part of America's underclass—the black migrant worker. It also celebrates the nobility of work and the spirit of survival inherent in these people. Of fundamental importance, also, is the relevance of research to locate, rescue, and record African American history and icons no matter where they are. As one Randolph resident said, "We are just tryin' to stay on the map."

In editing the interviews, I aimed to maintain the unique speech patterns, vocabulary, and voice of each narrator. One of my priorities was to let the narratives sound exactly like each individual. I want each person when reading the book to say, "Yeah, that sure sounds like me; that's the way I said it . . . that's me awright." Another priority was to keep the focus on the unique contributions of each participant within the larger, historical context. In the chapters on education, cotton, and women, for example, I place narratives that correspond with those areas of experience. Unrelated and tangential information is rarely used; neither are personal and confidential anecdotes about families, persons, and places. There were times, however, when I wanted to leave something out and an interviewee would say, "Oh, honey, I don't mind; it's the truth and part of my life anyhow . . . it don't bother me none." Although some of those things were interesting, I had to call on my ethical sense and make a decision, based on the "supposed" responses of their descendants in years to come.

In my introductions to each section and in the headnotes to each interview, I have aimed for a natural, simple vocabulary and style, free of heavy-handed academic jargon, so as to remain true to the class, ethnicity, race, and gender of the people being described and documented. The evocation of people, place, and moment is of utmost importance, and I wanted all to be vivid to the readers and the folk, my folk.

All of the people in all of those communities have touched my life, forever, in ways they will never know or truly comprehend.

I. Introduction

"There Are Pioneer Souls That Blaze a Path"

Dreamin' the West

> "I come to Randolph from Wellington, Texas, at the age of six years. Came in an ol' T Model. We had to take the top off of this ol' T Model before we could get over the hills. We came out in 1936. We had to push the car over the hills, it was in such bad shape." —Metris White

When comedian and actor Richard Pryor, in the 1980 film *Stir Crazy,* exclaimed that all the black people in Florence Prison of Florence, Arizona, "had to been bused in," he was echoing the impressions of many Americans about the absence of African Americans not only in Arizona, but throughout the American West. He and hundreds of people who have been miseducated about black American history simply do not know of the many contributions to Western lore, history, culture, and economy made by blacks over many decades. The fact that blacks were part of the Western frontier is no longer arguable; because of extensive scholarship undertaken in the last forty years and popular modes of presenting it, this history is now accessible to almost everyone.

More than eight thousand black cowboys helped drive cattle along the Chisholm Trail and experienced more equality there than they did in the towns of their origin, sharing with white cowboys the same rigors, rights, and punishments. Since the total number of cowboys in the West at their peak in the 1880s and 1890s is estimated at thirty-eight thousand, it is clear that as many as 20 percent were black.[1]

The westernizing of African Americans is exhibited in many ways, but none as romantic as contemporary cowboy Ed Keeylocko's dream to build a western town and "make a ranch that would support a town that would support a ranch," meaning that they would be interdependent. He did just that and named the town after himself—Keeylocko. It is located about forty-five miles southwest of Tucson, Arizona. As you enter this "semi-believable town," a sign greets you that reads, "Careful Stranger! You are entering the country where the Keeylock men roam the sweet sheree rim. It's the code of the West." By 1989, this town reportedly was recognized by its neighbors and had a mayor, citizens, its own zip code, and forty-six head of range-friendly, organically fed Brangus cattle, ten quarter horses, a long-horned goat named Harry, three ranch hands, 10,800 acres of land, five buildings, and one hundred mesquite stumps.[2] Many of the stories and myths about black cowboys were made into films or provided the content for nickel novels, a popular genre early in the twentieth century.

Added to these and other publications are numerous media presentations, such as public television's *The West* and *Tony Brown's Journal* documentary, "The Black West." Movies have contributed to the romantic notions of the black cowboy, the bad guy/hero, the Indian soul mate and scout, and the strong black man who spoke "Injun," who would cut down a white man in a minute, something he could not do in reality anywhere or anytime without his life being taken by gun or rope. Former athlete turned actor Woody Strode became the epitome of black machismo, a "colored" John Wayne and "the greatest Black western star, ever."[3] Some 1930s and 1940s films—for example, *Bronze Buckaroo* and *Harlem Rides the Range*—depicted the black singing cowboy. When these films were shown in eastern cities, black audiences laughed and applauded simply because these films had no relevance to life as African Americans knew and lived it. When more "relevant" films were made in the 1960s and 1970s, such as *Major Dundee, The Professionals,* and *Buck and the Preacher,* there was a better acceptance of them despite the fact that they were dubbed "black exploitation" films. Often the filmmakers were not knowledgeable about black western history and played up the romantic notion of what blacks could be in the West. One 1954 film, *Tomahawk,* featured the white actor Jack Oakie in blackface, playing Jim Beckwith, the legendary cowboy who always ordered two drinks—"one for himself and one for his horse." Then, of course, there was Nat Love (Deadwood Dick), around whom many legends of bravado developed. African Americans went to see these movies, often in segregated theaters, by the hundreds, and their fascination with the American West was no different from that of white Americans.

But the romanticizing was never enough to negate the contributions of African Americans as fierce warriors and clever, resourceful, and industrious builders. They were hardworking and resilient and contributed to the

Western economies of mining, farming, cattle ranching, and, in many ways, entrepreneurship. Irrespective of their participation, some facets of racial divisions in American life demanded that they be always "drawers of water and hewers of wood." However, it is only since midcentury years that scholars have begun to dig through old records and photographs and collect oral interviews to preserve a lost history. People are beginning to realize the value of such artifacts and of the voices of black Americans in the West.

This study, however, is not about the entire western movement of blacks; rather it is about those who eventually found their way to Arizona beginning in the late nineteenth century and continuing into the mid-twentieth century. These are the black Americans who probably had some sense of adventure, the chance takers who refused to see themselves as victims, had hardy spirits, and, to a large extent, displayed rugged independence. These were the black Americans who became victims of a western political economy that demanded that they stay behind the color curtain.

Well-researched studies of blacks in the West include William Loren Katz's *The Black West* (1971), *Black People Who Made the Old West* (1977), and *Black Women of the Old West* (1995). The life of the "Buffalo Soldiers," so much a part of Western military and African American history, is documented in William Leckie's *Buffalo Soldiers* (1967). In 1983, Richard Harris, an Arizona historian, published *The First 100 Years: A History of Arizona Blacks.* Thomas Cox's *Blacks in Topeka, Kansas* (1982) explores the early migration of blacks from the South into Kansas and other points West. The first blacks were a combination of military men, runaway slaves, scouts, individualists with pioneering instincts, explorers, cowboys, and others who contributed in myriad ways to the major western industries.

One of the earliest blacks to arrive in Arizona was the slave known as Esteban or Estevanico, who had accompanied Cabeza de Vaca from Mexico in 1526. When the Spaniards encountered the Indians and engaged in bloody confrontations with them, he escaped unharmed. He made his way through the treacherous mountain passes, unharmed by the superstitious Pima Indians, who were awed by his brilliant attire, feathers, and bells. Most reports about Esteban tell us that he was killed by the Zunis, but one report states that "Black Stephen" lived to be ninety-seven and fathered four children. Then there were the black mountain men, Jim Beckwith and Edward Rose. All three of these men now have monuments named for them—parks, towns, roadways, and schools.

The second group of blacks to arrive in Arizona consisted of many who were not cowboys, scouts, or free persons. Some came as maids, cooks, housekeepers, servants, or slaves to white families. This was also true in other western states such as Colorado, Nevada, and Utah. Utah records, for example, show that Mormon leader Brigham Young had a slave. Censuses and other records give varying figures as to the number of slaves in Arizona Territory by the 1860s. One census shows eighteen, which included three women; another shows twenty-six.[4] *The Sheriff* reported that 1868 saw the arrival of "the first member of the Negro race to reach the Salt River Valley." That "Negro" was Mary Green, a woman who accompanied the family of Lum Gray as a maid. Green had one child with her, a girl, and subsequently had five more. One of her children, Mose Green, is recorded as the "first child of Negro blood to be born in the Valley." By 1870, there were 127,000 Negroes in the West. Arizona had about 62; by 1880, it had about 155, and by 1890, 1,357.[5] By 1900, Phoenix had 200, and Tucson, 90. That number grew with intermarriages, but by 1912 Arizona's miscegenation laws forbade intermarriage; they were not repealed until the 1950s.[6]

The presence of blacks in some western towns was accepted and tolerated because, "in those years, a great many of the residents . . . were of southern extraction or connections and the local Negro folks were kept strictly in their place, and, at that time, caused no trouble or disturbances of any kind—either socially or commercially."[7] Despite this general mood, things began to change by the 1920s when the Ku Klux Klan openly paraded in Phoenix. By this time, the black communities in Arizona had family units in place, with established churches and schools. There was also a certain amount of jealousy on the part of the also newly established white populace, who were already sharing homesteads and other necessities with Mexicans and Indians.[8] The oppressions and enemies the blacks had faced in the South and the other places from which they had come were present in the West as well.

The third group of blacks to migrate West were the escapees from oppression and the determined ones who were part of the "Exodus of 1879." They came from the former southern slave states to flee the disruption that took place during and after the Reconstruction era of 1865–1877. As black people again began to experience the same horrible incidents they thought had ended with Emancipation, appeals were made to the federal government in Washington. But nothing was done. A committee headed by Henry Adams and Benjamin "Pap" Singleton came up with the idea of leaving the South and going to Kansas, as well as other nonslave states west of the Mississippi River. William Katz wrote:

> Henry Adams and his committee organized what was later called "The Exodus of 1879," a huge black migration into Kansas and points west.

> In one year, an estimated 20,000 to 40,000 penniless and ragged black men, women, and children reached Kansas. They came up the Mississippi by river boats, or made the slow walk up the Chisholm Trail. They were spurred by the knowledge that it was there in Kansas that John Brown first struck against slavery. And they hoped there to own land and homes.[9]

White southerners reacted with fear and panic at the thought of losing their cheap labor supply. Black leaders of "the exodus" were denounced as troublemakers, and local agents for "the exodus" were driven out of town or beaten. By May 1870, white Mississippians had closed the river and threatened to sink all boats carrying black migrants. Gen. Thomas Conway, in a letter to President Hayes, reported: "Every river landing is blockaded by white enemies of the colored exodus; some of whom are mounted and armed, as if we are at war." One Kansas settler who came back to get his family was seized by whites who cut off both his hands and threw them in his wife's lap, saying, "Now go to Kansas to work!"[10]

This group's hardships and journey are well documented, and the people are usually referred to as the "Exodusters." It is estimated that between twenty thousand and forty thousand blacks moving from the South left for Kansas and the western states and, they hoped, freedom. Benjamin "Pap" Singleton "viewed the emigration to Kansas as the embodiment of racial self-confidence. . . . a way to get away from white mob violence as well as race-related civil and economic disorder in the South." Singleton is credited as the instigator of the Black Exodus. John Mercer Langston, a prominent politician at the time, viewed this mass migration as "an expression of the freed man's exigent demand for independence without which no individual and no people can rise to the level of dignified and honorable manhood."[11] Many remained in Kansas, endured hardships, and created black towns such as Nicodemus. Few returned to the South. Hundreds moved into Oklahoma, where the harsh land, hostile people, and poor farming discouraged them daily. In Oklahoma they again created towns and communities, such as Boley and Langston. Flyers were distributed nationwide inviting all "colored CITIZENS of the UNITED STATES" to discover freedom and invest in these new places, to buy land, build, and be treated honestly and fairly. Nicodemus, Kansas, and Boley, Oklahoma, still exist today as black towns and "settlements."

Black Americans did not come to the West in great numbers to work until the late 1930s. Early black pioneer families or single persons came to escape oppression, to look for land, to homestead, and for health reasons.

But the Black presence never received public recognition because, despite the "open West," the wretched hand of racism placed limits on how far black settlers could go. There are many stories of swindles in Western lore, and the groups constantly in the grip of white American injustice were blacks, Native Americans, and Mexicans. But some mining companies incorporated with the help of black Americans. Henry O. Flipper, a black cavalryman, worked for Colonel Green, a white man. Other men and women did mining work in the Bradshaw Mountains and in Tombstone, Bisbee, Wickenburg, and Morenci. Caleb Martin, an ex-slave, headed one of the largest and most successful ranches in Arizona. His children became cowboys, rodeo riders, ranchers, and ranch cooks. In the cotton industry, those blacks who were recruited to pick cotton seasonally settled in towns such as Eloy, Marana, Randolph, and Somerton. Henry Foley knew how to do more than just pick cotton; he was paid more to "press the seed," which means he worked in the cotton gin, a place usually reserved for white workers.

Yet none of the three major industries in Arizona— copper, cattle, and cotton, called the "three C's"—involves black Americans in any major way at the present time. The copper industry has had many financial setbacks and has developed more modern ways of mining. The cattle industry never had major black ranchers except for Caleb Martin in Wilcox. And as cotton picking became mechanized, workers without other skills were left destitute or became itinerant seasonal workers in California and other western states.

The fourth group of black people moving further West came under different circumstances. They left places such as Arkansas, Texas, and Oklahoma to look for work, food, shelter, and perhaps "paradise." A breakdown in the western economy began by the late 1940s and early 1950s, and it affected not just black Americans but white Americans as well. Miners, ranchers, and farmers had to realistically realign those industries. The preceding decades had brought the devastation of the Midwest by natural disasters, including droughts and floods, as well as the Depression of the 1930s. Nature, the great giver and taker of life, challenged this population of Americans to survive or pass on. But their spirit and determination brought a great mass of humankind together to strike out for a new life filled with hope and dreams for betterment and new beginnings. While whites set out for the West primarily to work, blacks were seeking not only to work but also to escape segregation and all forms of racism, which seemed to have escalated as the economy began failing. The romantic notion of the West had changed, but even as their forebears came in wagon trains and on foot to seek freedom and realize their potential, they were coming to seek work and begin again.

During the 1930s, if you were white and from Oklahoma or Arkansas, you were called an "Okie," a term meant to be demeaning and dirty, a term assigned to those outside the class lines clearly drawn by middle- and upper-class whites. If you were black and also came from those states, as well as from Kansas, Missouri, or the farming areas of Texas, you were classed below the "Okies," because whiteness still brought special privileges no matter the class. Both groups suffered similarly, but also in many ways differently because "not all okies were white."[12]

In the 1930s, 1940s, and early 1950s, Route 60 became the black migrant road, just as Route 66 was the white migrant highway. As the Depression years hung on, seemingly forever, the conversation of travelers on the highways was about where they had come from, what and who they were, and where they were going. "I'm from Oklahoma—Cherokee, Salisaw, Checotha, Hugo." "Got us a car in Oklahoma City, had a store there; going to Arizona to pick cotton." "Goin' to Barstow to pick fruits and vegetables." "We're from Poplar Bluff, Missouri, came through Joplin, folks was 'croppers, the strike of '39 put us out on the road." "We're from Arkansas, Hope, father died, mother takin' us West." "See these handbills, plenty of work in Arizona, good clean camps, big cotton harvests." "There're seventeen of us in this trailer, got to get there by Friday. Uncle who came two months ago, waitin' for us to pick; Dad will be here by December." "Got no food or money, got to work our way acrost." "Had land in East Texas, Marita, crop failed, that ol' Mother Nature turnt it all upside down." "Was workin' a crop for a white man. Put a gun to ma head and say we couldn't leave 'til we picked all of it, just like slave days. Me and the wife and kids had to push that ol' car in dead a night to get away." "Come in last night from Paris, Texas, got a brother in Tyler, another in Waco." "Some people are turnin' back, say some of those flyers seen in Oklahoma City is full a lies, but we aim to keep movin', got to keep movin'!" The voices of blacks and whites blended on those highways. Both had similar dreams and hopes. All were leaving the worst behind them.

As these migrants moved along both highways—one group on 60, the other on 66—people in the towns and at the road stops passed judgment without emotion. "If that ol' jalopy makes it, it will be a miracle out a scripture." "They ain't human. Humans wouldn't live the way they do." Along Route 60, which many joined in southern Missouri, at places like Sikeston and Springfield, on down to Oklahoma, across northeast Texas, and through Clovis, New Mexico, passing into Arizona, and through places such as Globe and Superior, down into Apache Junction and west to Phoenix into the farming areas of Glendale and Vicksburg, they rode on; sometimes "workin' their way acrost."

Courtesy of the Jordan family

Black migrants en route to Arizona.

In the Casa Grande Valley with its hundreds of acres of cotton, black families came to pick. They came to Eloy, Chandler, Buckeye, Cashion, Casa Grande, Coolidge, and Randolph. They came not only to pick and make money but also to start over, to find out. "There's something going on out there and I aim to find out what it is," said Casey in John Steinbeck's 1939 novel, *The Grapes of Wrath.* "This is my country, my dirt . . . ain't no good but it's mine; I ain't a goin'," said Muley in the same novel. These two very distinct messages uttered by Steinbeck's white Okies were similar to ones made by the black Okies in the 1930s and 1940s.

One group, the white, was responding to the devastation of nature's great dust storms, floods, and the crop damage from a little insect called the boll weevil, which attacked cotton from Mexico to California. The other group, the black, the underclass, the poorly educated "working hands," also sought land where they could work. A lot had changed since Pap Singleton, Frederick Douglass, and Booker T. Washington and those nickel novels and movies. Migrants from both groups left Missouri, Arkansas, Kansas, Texas, and Oklahoma looking for work and a place to stay. Little by little, the downtowns

Courtesy National Archives

Market day in Eloy.

in Oklahoma and Arkansas that boasted of barbershops, drugstores, movie houses, and many businesses along main thoroughfares started to die—some quickly, some slowly. Like those other Okies, black families began looking West with hopeful eyes and determined hearts, to make it. The hundreds of miles of Routes 60 and 66 witnessed a human ecological movement westward, crawling along highways and rural roads to the unknown. The fact that they followed different routes is significant in its reflection of the extent to which America has inherently incorporated into its culture markers that denote difference, even in travel.

The history of black migrant movements for opportunity is filled with dashed hopes, mental lynchings, treacherous acts, unscrupulous dealings—without punishments to the perpetrators. Often there is simply silence. This project incorporates some of the stories the survivors tell about themselves, those migrant journeys, and the final destinations. For some, to have survived is close to miraculous.

This study is a first in that it focuses on the black cotton pickers on the very productive farms in the Casa Grande Valley between Phoenix and Tucson. It is the only study, to date, that uses oral history as the primary mode for conveying information about one small black town in Arizona. Although other towns have similar stories, Randolph is the major focus here because it is an anomaly in the twentieth century, embracing many facets of black life and its impact on other communities and peoples. Unlike Dearfield, Colorado, founded in 1910 and "once a thriving agricultural colony comprising more than sixty families . . . and now a ghost town,"[13] Randolph has remained even after the cotton work was taken from it and the black population dwindled. It has put into practice Booker T. Washington's lesson that the best hope for blacks lay in self-help and economic independence. Many residents sought to achieve just that, often eking out a living under meager circumstances. They are the real and new heroes of Western history because, after the machines came and replaced workin' hands with iron claws, work and opportunities diminished, class lines tightened, society became more stratified, and environmental racism further destroyed their tired bodies, yet they stayed to rebuild their lives. Theirs is a testament to the black spirit of overcoming and movin' on.

II. “Men That Press On with the Ardor of Hope”

Migrant Places and Randolph, Arizona

> "I think they were brave to come here. . . . There's no tellin' what was told 'em back there where they came from." –Sarah Lucille Smith

In the 1920s, white Arizona speculators took advantage of as many suckers as they could—black and white. There are many stories of land swindles, such as that of Bouse Wash, or "Nigger Flats," near Vicksburg in 1925. Black people from Los Angeles were promised cheap land (320 acres of "good" farmland at a price of $1.25 per acre and a down payment of $50.00), precious well water, and work at the nearby mines and ranches. It was too good an opportunity to miss. The offer was accepted by good churchgoing black folks who believed that life would be better there than in the towns and cities that had already disappointed them. Not only did the swindlers skip town, but the blacks were assigned to the flat lower section of the area, which would flood whenever it rained—thus the name Nigger Flats. It was located on U.S. Highway 60 between Hope and Brenda, and according to Richard Harris "was to be something of a utopia for some 500 'Pilgrims' from Los Angeles."[1]

Nigger Flats is one of three black communities in Arizona that arose from a desire to resettle and gain permanency and a sense of self-contentment that had been elusive in other unfriendly states. The second is Mobile, south of Phoenix and named after Mobile, Alabama, where a group of blacks purchased large amounts of land only to find their water rights inadequate. The fifty-eight families who migrated to Mobile, Arizona, bought into the plan of a white preacher, named Blevin, to create a religious community. The community was dissolved after Blevin was sent to prison for unsavory investing and promotions. Mobile still has a small but scattered black population, as many families from Phoenix and Casa Grande, seeking quieter, safer places to raise families or to farm, settled there. The Mobile Elementary School is pretty much what now marks it as a town. Various ventures were tried in Mobile, and one, a hog farm, did pretty well. Mobile remains, like others, an unincorporated town in Maricopa County with few services and is now the recipient of toxic dumping from Arizona and California.[2]

The third town is McNary, Arizona, named for James G. McNary, who was the co-owner of a McNary, Louisiana, lumber mill. After the forests around the mill in Louisiana became depleted, he and his partners, William Cady and Alfred Smith, left to find new forests. They came to the Apache Indian Reservation area in the White Mountains, which was beautiful and lush. It was called Cluff Cienega, and later renamed Cooley for Corydon E. Cooley, an army scout. Their Apache Lumber Company employed "400 men and could manufacture 175,000 board feet of lumber a day." When the owners got in trouble financially, Cady and McNary were opportunists and purchased the company in 1923. James McNary in his later years wrote, "Cady could not visualize a lumber operation without the employment of black labor, and he decided to import about 500 of our experienced and faithful [Louisiana] employees to Arizona."[3]

In 1924, with the promise of steady work, good living conditions, and great weather, seven hundred hopeful black migrants left Louisiana for northern Arizona. They carried everything they owned with them—household items, beds, animals, chickens, and their memories in trunks and sacks. But many, after finding that McNary was in the cold, snowy northern mountains, returned to Louisiana, discouraged. "The town grew with a Negro quarters and a Spanish-American quarters, each of which had its own elementary school, church, and café. There was also a small gathering of Navajo bush cutter's shacks and a gradually increasing Apache town over the line to the west."[4] Thus minorities were segregated not only from whites but also from each other. Black families lived in an area called "The Hill," which was once a neighborhood for whites. In 1990, black resident Vera Blake said that there were about twelve or thirteen black families remaining on "The Hill," along with some Hispanics and Native Americans who stayed in their "Quarters."

Then there's Randolph. Unlike the black towns of Oklahoma and Kansas, Randolph did not begin as a black town but was advertised as an area in the Casa Grande Valley that would rival Phoenix in industry, growth, and population. It was a project of white speculators of the Randolph Townsite Company. An article in the *Arizona Republic* in 1925, entitled "The Romance of Randolph," speaks glowingly of Randolph's potential. It describes southern Arizona as the "gold spot of the country" with a "great area of irrigable land, with great canals winding here and there placing water where it is needed . . .

green alfalfa fields—full bloom cotton crops, miles of cars carrying cantaloupes to Eastern markets . . . five to seven crops of cotton—the cotton, a quality that brings the highest price on the market . . . a main line railroad, a national highway."[5] This detailed and overblown article speaks of Randolph as a place to participate in the American Dream, which will be a "romance of potential growth undisguised." Randolph did not develop as planned but became an unincorporated town in Pinal County. It is named for Col. Epes Randolph, who was vice president and general manager of Southern Pacific Lines in Arizona. Perhaps the site, as a station on the Eastern railroad line, was seen by Colonel Randolph and his partners as having the potential for development. Instead, the town was settled by cotton farmers in the 1920s. Will C. Barnes in *Arizona Place Names* describes it as follows: "A Randolph Townsite Company in Phoenix advertised Randolph as being in the heart of the San Carlos Irrigation District on the Southern Pacific. It was located fourteen miles southwest of Florence, sixteen miles northeast of Casa Grande on the main line railroad in the very heart of the San Carlos Project. Composed largely of farmers and landowners, it was thought that Randolph could only go forward, that growth was inevitable."[6] The agricultural boom of the late 1920s created a need for day and seasonal workers, to pick and chop (weed) the thousands of acres of cotton in the valley; thus the migrant roads led hundreds of pickers to the area and its surrounding communities.

The black migrants who came into Arizona and dispersed in various farming areas, where work was to be found, were the children and grandchildren of the "Exodus of 1879" and the great migration of 1918. Their obituaries and oral autobiographies show birth dates from the late 1880s and 1890s on into the 1910s and 1920s. They were the generation who created these towns and stayed in them—Eloy, Casa Grande, Vicksburg, McNary, Cashion, Mobile, Randolph.

The presence of blacks in Randolph dates back to the 1930s, when cotton farming first became profitable in southern Arizona. Black migrant workers from Arkansas, Oklahoma, and Texas began to hear of the employment opportunities in this new Cotton Belt and came to work on the expanding farms during the harvest season from September to December. One resident said, "They got us here by telling us there was money on the ground and all you had to do was rake it up and put it in a sack." Of course, that was not the case. That was part of the Western lure of opportunity, enchantment, work, promise, and untruths, which has permeated American history for decades.

Randolph has a unique history as a town, because it started out not as black, Native American, or Hispanic, but white. Because of labor needs, geography, economics, and culture, it became a different town and an experiment in the coexistence of people and the human spirit. In its early days, Randolph was 100 percent white, averagely populated, and surrounded by cotton farms and family businesses. When one drives into Randolph on Highway 87, which runs north from I-10 or south from Coolidge, what one notices is that the highway divides the town and that one side, the west, has white people and the other, the east, is black, with a few Mexicans on the periphery of the east side. Blacks and whites everywhere have for decades been separated by highways, freeways, railroad tracks, or some other physical and discernible border. However, in Randolph, the county and town records of 1925 specify that "All resident lots west of the railroad are restricted to those of the white or Caucasian race." Further, the records show that particular choice lots were designated as "for whites only," while others on the east side were for those of "the non-Caucasian race." The price asked for each lot varied from $125.00 to $250.00 generally. The highest were a few for $265.00 each.[7] Additionally, the building plans further stated that "all industrial plants were to be restricted to lots east of the railroad." This, of course, is the black side of Highway 87. Those industrial plants were indeed built on that side of the town, and more and more of them have been built in recent years. So, the residents suffered not only from racism born of differences in skin color, class, and economics, but also from geographic racism. Today, when we speak of "environmental racism," we can see that it is not new, and one reason so many Randolphians suffer from lung and a variety of other health problems is their long years of exposure to the unhealthy surroundings coupled with years of hard physical labor.

One black resident of Randolph referred to it as being "just like an Indian Reservation." This is an appropriate description because not far from Randolph is the Gila River Indian Reservation, where thousands of Pima Indians were "banished" or set aside just as blacks, historically, have been. Randolph's black residents also picked with the Indians, their cousins of color, and a few intermarried with them. One sees that mixture in their faces.

A brief history of white Randolph begins about July 18, 1925, the date Channing E. Babbitt was named the first postmaster, under Gov. George Hunt. As postmaster, Babbitt was the official keeper of the town's records. His monthly log of stamp sales shows that stamps cost two cents at that time and the total intake for most months at the Randolph Post Office was between twenty-eight and fifty dollars. The mail was picked up by rail; Babbitt would set the mail by the tracks or throw it to a train engineer. The residents in those days actually believed that Randolph, with its romantic advertising, the Townsite Company's dream, and the promise of a growing metropolis, would become

Roof's Welding and Machine Shop.

a city to rival Phoenix. Their hopes were shared with the black laborers who arrived a few years later. The white presence in Randolph can be cataloged, but each family name or event has its own story, reminiscent of a southern Faulknerian hamlet.[8]

Babbitt, who came from New York, married a cousin of May Smith. May came to Randolph in 1927. Babbitt also had a wood hauling business, an icehouse, and a service station. The western spirit of enterprise was not lost for this easterner. May Smith became postmaster in the 1940s and died in 1954. Babbitt died in 1957 at the age of eighty-five. Sarah Lucille Smith was the last postmaster, from 1955, the year a new post office was built. She remained postmaster until 1967, when she retired.

Leon Thompson and his father and mother, E. O. and Cappie Thompson, had a service station and bar. Clifford Griffin and his wife, Gertie, had the Randolph Market. Gertie was a daughter of E. O. and Cappie. A. E. Woods and his family had a grocery store. Two of his sons, Red and Esper, and their families lived in the town. The Marshall brothers, Vernie and Manuel, both veterans, opened a Texaco station after the World War II. Their parents lived in Randolph also. Ralph Veazy had a welding shop. Calvin Roof and wife, Gussie, had the blacksmith shop north of the welding shop. Gussie still resides in Randolph. Her parents and Dave Hughes were in the hay-hauling business. Their son Wayne still has the shop. Connie, Gussie's daughter, was Miss Pinal County. Steve and Ruth Soule had a farm in Randolph in 1925. Later they opened a hardware store. Clyde and Edna Monday bought the Steve Soule property and kept the store open.

The Quintero family lived at the store. They're Yaqui Indians. The father, Augustine, and son, Manuel, had tuberculosis, and Manuel had one lung removed. One daughter had tuberculosis of the bone. A granddaughter now lives in the old home.

The Youtseys, an old family, worked with cattle and hay. They're Mexican. Some other old ones were the Hollands. The Loopers and Acunas were laborers and worked with cotton.

The Arizona Training Center was built in 1953.

The post office closed in 1983.

Water was hauled until Mr. Carter drilled a well and the Arizona Water Company bought it. Lee Smith owned a well that served twelve families with piped water. Some of those names are long gone, but some linger with great-greats who no longer live in Randolph. Their spirits hover on the west side of Highway 87.

The present Arizona Training Center, where many Randolphians, black and white, work, was at one time the Arizona State School for Girls and, after that, the Children's Colony. The State School for Girls, which opened in 1928, was a detention school for "bad" girls and had a long and unpleasant history with the state. Its mission was questioned, its placement questioned, its finances questioned; there were all kinds of investigations about abuse, laxness in its operation, lack of supervision, and poor management. An October 1931 report, prepared by a Mrs. Lavers, is thorough about the causes of delinquency, the school's mission, and other statistics. The school closed in 1935 under much scrutiny, with much controversy, and bringing an end to many careers. It ought to be the subject of further study. There was also an Arizona State School for Boys at Fort Grant, which suffered a similar fate. Renamed the Children's Colony in the 1950s, the Randolph facility began to fare better, serving a population handicapped emotionally and physically. This evolved into the present Arizona Training Center. Some say that name sounds better but is still inappropriate, since the facility houses mainly mentally handicapped persons. The institution is about one-half mile from Randolph, on the west, or white, side of Highway 87.

At the town's peak, with many businesses flourishing, the cotton crops producing, black, Mexican, and Native American pickers pickin', the Randolph whites and some of the blacks were doing well. Most of the white businesses, the grocery stores, gas stations, auto shops, icehouse, and post office, which primarily sat on the east side along Highway 87, flourished from the business of the local laboring class and the constant flow of traffic along the roads.

The *Coolidge Examiner* in the 1940s had a society and gossip column of sorts called "Randolph and La Palma," written by Katherine Hupfel. In it she related all of the social gatherings at various white homes, including dinners,

The dedication of Randolph's new water tower, in the early 1960s.

parties, the arrival of guests from out of town, anniversaries, and other items of popular interest. For example, she wrote:

> "Bud" Woodruff of Tucson, spent the weekend visiting at the home of his aunt and uncle, Mr. and Mrs. E. O. Thompson; Mr and Mrs. Warren Steele, Jr. and son recently have moved into their new home on the Steele Ranch south of Randolph; Mr. and Mrs. Lee Smith and family had dinner guests at their home in Randolph Christmas Day.[9]

Sometimes Hupfel would include small items on Mexican or black residents. Often these were not those considered "social news":

> Seen on Main Street Saturday. . . . A black gal was walking along minding her own business and looking straight ahead. She was quite personable, as black gals go, which didn't escape the attention of a "dressed up" nigger strolling in the same direction on the opposite side of the street. . . . "Wha' you goin' to?" he called. . . . "I'se goin' wha' I'se lookin'," she answered sedately. . . . He considered this. "If you is goin' wha' you's lookin', what fo' you don' look ovah heah?" he inquired persuasively. . . . She turned her head slowly and surveyed him with evident approval. Then, still looking "straight ahead," stepped down from the curb and crossed to where he was. . . . Grinning broadly, he linked his arm in hers . . . a couple of black birds, "going where they were looking."[10]

Several weekly columns by Hupfel, like this one, could be considered racist. Other columns only reported "Negro" crimes such as drunkenness or fatal accidents. Randolph and its surrounding towns were comfortable as long as blacks stayed in their place on the other side. This was also a time in America before the advent of the slogan "Black Is Beautiful," the Civil Rights Movement, and the right to be called "Afro-American," rather than a colored boy or gal. There was no NAACP chapter in the community to fight such things. The closest association would be in Eloy, later on.

Not much has been written about black towns in Arizona. Katz's *Black West* and Leckie's *Buffalo Soldiers* do not deal with the subject. The West in those texts is Oklahoma, Nevada, Colorado, California. Harris does mention Randolph and McNary in *The First 100 Years: A History of Arizona Blacks.* We know of black Moor Estevanico the explorer, blacks as victims of land swindles such as the "Nigger Flats" episode on Highway 60 near Vicksburg, as well as the disillusionment of the lumber mill workers of McNary, lured from Louisiana to create a life in warm, sunny Arizona. Randolph, Arizona, is clearly, then, an anomaly in the West, and only one of a handful of black towns that remain in the United States. N. Y. Nathiri's *Zora* (1991) is based on the life of novelist, anthropologist, and folklorist Zora Neale Hurston of Eatonville, Florida, one of America's oldest black towns and the first black incorporated town. But Zora's town and name are well known by most academicians and others. The Negro migrant workers in Randolph are not so fortunate.

Officially, Randolph is not a town at all, but an unincorporated part of Pinal county. It has no mayor, no governing body, no police, and no direct services except for electricity, water, and telephones. Garbage is taken to a dump site eight miles away in Florence, the county seat. Some mail is brought from Casa Grande, about thirteen miles away, and put in streetside boxes, but many residents have boxes in Coolidge, three miles away. Even though the streets of Randolph have names, the town name in each address is either Casa Grande or Coolidge. For example: Mr. and Mrs. Brown, 100 Martin Luther King Street, Casa Grande, Arizona 85722 (Casa Grande's zip code), even though King Street is in Randolph.

In the 1940s and early 1950s, black migrants began to make their presence known in Randolph. Many bought lots on the east side of Highway 87 on which they built homes for themselves and other family members. Nicholas White Sr. not only owned several lots but also had a small grocery store. Nick Jr. had a Texaco gas station, which he said "did pretty good." He speaks candidly about the kindness of a white man who cosigned so that he could get the business started. Other black-owned businesses along Highway 87 included a nightclub and bar. Some Saturday nights were very noisy as blacks made their own good-

The camp housing provided for migrant laborers, as it looks today.

time parties after a week of hard work. One said, “We always had dances on the weekends, and we kids and our friends played around the camps.”

As the new migrants would discover, the whites owned everything, while the blacks and a significant number of Native Americans and Mexicans were the workers. White farm owners housed the workers in small shacks, paying them three cents for each pound of cotton they picked. Some of those shacks can still be seen (now ruined) as far away as the freeway around Florence, Eloy, and Casa Grande. At its peak in the 1940s, 1950s, and early 1960s, Randolph had a black population of approximately five hundred. At that time there were some black-owned farms, gas stations, cafés, and other businesses. The Night Owl bar was a local hangout, and the Galloping Goose sign between Coolidge and Randolph really “galloped.” Residents tell of walking up to that bar just to see the animated flickering lights. Randolph even had an elementary school for black children. That is no longer true.

Beulah Moore, who came to Randolph with her children from Hope, Arkansas, in the 1940s, had a small contracting business. She used her one truck to drive pickers to the fields of neighboring farms. She was perhaps the only woman who had her own contracting business at the time. It wasn’t big,

The Galloping Goose bar has been operating at the south end of Coolidge since the 1950s.

but it allowed her to buy quite a few lots and raise three sons and a daughter. The youngest son, Melvin, would become a successful professional in law enforcement as a Pinal County deputy sheriff, and some consider him the unofficial mayor of Randolph. His son Alvin attended Arizona State University as an athlete and subsequently played football in the National Football League.

Another labor contractor from Hope, Arkansas, was Cleveland "Preacher" Jordan, whose father preceded him to Randolph. Jordan had four trucks, which he used to make trips back and forth to Arkansas to recruit pickers. The Jordans had their own labor camp on their substantial landholdings a little southeast of the center of Randolph, near the railroad tracks. Cleveland and his wife, Myrtle, a Pima Indian, had a food business that the migrants depended on for meals on credit. He also hauled water for the workers and used his trucks to transport them not just to the fields but also to town to do their weekly shopping. The Jordans still own much of that property.

The Obed C. (O. C.) Fields family came to Randolph not to pick but to buy land and farm. They had been to California after leaving Falls County, Texas, but did not care for the California cities, so they came to Arizona's Casa Grande Valley, where they acquired 180 acres of land for cotton farming. The entire family worked this land, and they still work the forty acres they now own. Two Fields sons are in their late seventies; O. C. is ninety-nine. They raised not only cotton, early on, but also alfalfa, cantaloupe, and watermelon. O. C. is probably one of Randolph's most successful "migrants." The Fields place became a center for people from the surrounding communities of Eloy, Coolidge, and Casa Grande to have barbecues, church celebrations, and other activities, until the mid-1960s, when things became quieter. The big picking machines had come, handpicking declined, and people began to leave those communities or had little to celebrate.

Then there are the Dostys, who are related to the Whites; Robbie Dosty became a basketball star at the University of Arizona. And the ever-present, large Muldrow family, whom everyone says you can identify anywhere because of their light skin and freckles; the Glasses (one married a Reece); the other Fields who married a Selmon; the Selmon who married a Moore; the large Jordan family, whose members intermarried with all races. Then there are those strong determined men like Jack Banks, farmer, Coolidge police officer, prison guard, market owner, army officer, and preacher, who said, "When you grow up in Randolph you just don't have what other kids have." And Gene "Duke" Thurman, who used to drive the schoolbus from Randolph to Coolidge; Clayton Rushing, whose former teacher at Eleven Mile Corner School said he was "good with his hands, and smart as a whip"; and Garret Langston Turner, called fondly "GL," who owned several lots in Randolph but never really developed them, although he did sell "stuff" at the swap meet every week; GL, a short, friendly man who always wore a wool cap, summer or winter.

Randolph's women are special, determined, and permanent, and their strong presence is felt everywhere. They are the caretakers of everyone and everything. They are held in a special place of respect in Randolph because it, like them, has seen many "women's times"—good and bad.

Courtesy of the *Arizona Daily Star*

Melvin Moore.

Clara Bell, the midwife who assisted so many births for so little money, did so because her "sisters" couldn't afford any more, or could not be admitted to regular hospitals because of their race. This mother of all mothers had a street named for her in the 1960s.

Fannie Reece, fondly called Ms. Fannie, who came from Hugo, Oklahoma, in 1950, "on the truck," has become the town's matriarch. She mothered nine children and grandmothered and great-grandmothered several more. Her black and Cherokee features are "keen," and her eyes twinkle as you enter her presence; she has no harsh words for anyone. One son, L. B. Reece, became a Golden Gloves boxing champion and fought in Madison Square Garden in New York City. She speaks of that proudly. He was Randolph's celebrity for a time.

Mamie Selmon, a fiery, feisty elderly widow from Oklahoma, did not work "in the cotton" but did some day-care work. Mamie never took the negative words or actions of whites without fighting back. In her words, she "would slap one of them in a minute." Mamie, who remembers such scenes in Oklahoma, still emits an anger and strength as she chews her snuff and watches out for grandchildren. Her words are emphatic, anguished, still tinged by old angers. One daughter, Mandy, married a Moore; another married a Fields.

Metris White, whose father, a Dosty, came to Randolph as a widower from Wellington, Texas, herself married a White—Nick. Metris is the official "keeper" of Randolph's population records—births, deaths, and departures. She says, "I can tell you how many left here, just by countin'." Her obituary collection is a tribute to her town. Metris, an early arriver, came as a very young girl in 1936.

And, of course, there is Lenora Mathis, sister of the Reverend Nick White and mother to ten, who moved to Randolph in 1955. And Zelli Thomas, one of the first blacks to come to Randolph, who arrived in 1938 and never left. There are many others from those other places in Arkansas, Oklahoma, Texas—many, many from Texas. Mostly, the women are widowed, but never alone. None forget what they have been, and all take pride in the fact that they have survived even the machines to raise another generation of nonpickers.

There are two churches in Randolph—one is St. Paul Church of God in Christ at King and Bell Streets. Rev. Nick White has been its pastor since he returned from World War II and the Pacific. It burned down once but was rebuilt with caring hands. The other is the Randolph Community Baptist Church, whose minister does not live in Randolph. Neither lacks parishioners, and both supply the strong religious base that is so much a part of black life. They administer to all. There are no factional battles about who belongs to which congregation or which preacher gets 'em goin' best on Sunday mornings. These are a moral and God-fearing people, full of tolerance, who know no malice.

Rather than developing, Randolph is losing population and visibility. The cotton industry's growing automation since the late 1950s "took the money right out of people's pockets." The one convenience grocery store, owned by the Melvin Moore family, closed in 1995. Randolph now is mostly black, with the exception of about twenty residents. These are primarily the Mexicans who move in periodically and a few whites who still live on the west side of Highway 87.

The St. Paul Church of God in Christ.

Some residents complain that segregation helped Randolph and that integration in the mid-1950s destroyed it; that is, with integration came less need for black businesses and work. Unemployment is high, with many between the ages of sixteen and forty-five—the most productive years of one's life—out of work. The median income per year for a family of four is $8,000, below the Arizona Department of Economic Recovery poverty level as of 1990. Yet, the black revolution of the 1960s and early 1970s touched Randolph despite its isolation. The VISTA (Volunteers In Service to America) social uplift program of the Johnson presidential years provided help and outreach for the community. The young and idealistic men and women (primarily white) of VISTA were welcomed into the community and were missed when their tenures were completed. One of the uniquenesses of the town is its streets named for Martin Luther King Jr., Malcolm X, Stokely Carmichael, and John F. Kennedy—heroes of the Civil Rights era. Rev. Nick White said, "The young folks got together and picked out the names." This naming took place during the VISTA years.

As the economy of Randolph changed and declined for blacks, it did as well for whites. Businesses such as the Randolph Market closed, the icehouse stopped delivering ice, the gasoline stations closed, as well as the blacksmith shop and the appliance store. The only white business that remains today is the Roofs' machine shop, which is run by the male descendants of that

The Randolph Market closed when hard times came.

family, but it is quiet for a machine shop. Whites still live on the west side of Highway 87 as they did seventy years ago, but it is not unusual to encounter a white child who goes to the black side to play with a schoolmate or visit a neighbor.

The houses on the white west side are not fancy; they too show the signs of aging. The yards are not manicured or splendid. There are no lawns or gardens. Most live in trailers surrounded by the usual old collections, which are kept too long and should have been discarded years ago. These houses are no different from those on the "colored" east side. It seems as though they are one and the same, similar to those of the 1930s, 1940s, and 1950s. A driver going through Randolph, looking left or right, would note no difference. Blacks and whites share this same slow decay. Some of these Randolphians are the Okies of Route 66 who now live similarly to those black Okies of Route 60 across the highway. Both look at boarded-up businesses and feel a strange quiet in the Arizona desert.

Randolph now has a small and dwindling population, and 90 percent of them are part of America's underclass. When "King Cotton" died and the machines took over, some stayed, and some, like the "Okies," moved on. Most of those who stayed are women, and the elders and the town reflect that very strongly.

Randolph's history for blacks was very much connected to cotton in the peak years and afterward, and cotton was the crop that influenced the development of the community. One resident said: "When I moved here in 1940, they [the whites] owned everything, but just like that, blacks started staying and building their homes." But as machinery took over and hand picking stopped, life slowed for blacks by the 1960s. The main sources of employment now are the nearby Arizona Training Center and businesses in the towns of Coolidge, Florence, or Casa Grande. All are difficult to get to because of the lack of public transportation from Randolph.

Some older citizens reminisce about the importance of water for the little community back then. "Once we couldn't never git no water worthwhile. We had to pay two dollars a barrel for it to be hauled to our homes. But folks got tired of that and started leaving Randolph." "Ain't much of a town no more, 'cept old folks and children. Soon as they grow up, they gone."

Randolph residents have differing images of the town. The elders speak of how things were and that it wasn't all that bad, that life had a purpose, honest work was respected, and family and community closeness were valued. The younger ones show frustration and impatience with the poor educational and job opportunities. One older citizen said: "I remember the days when people lived on cotton camps and went to chop [weed] when it was dark and came back from chopping cotton when it was dark and we did good, we didn't worry none." Despite the varying opinions, the overall feeling is "that people care about each other and take care of each other . . . a lot of people think that we're starving to death, but we're probably doing better than a lot of people in Tucson or Phoenix, where you get ripped off if you go out for a soda pop . . . here you can spend a week in California, leave your house unlocked, and nothing happens."

There is a feeling in Randolph of everyone caring for and taking care of everyone else. Children walk and ride bikes in the center of the streets and are in no danger. They play freely and visit each other's homes all day when not attending the nearby Coolidge Public Schools. It is not unusual to have several related families in town, a father and mother, their children, their grandchildren, cousins, sisters, aunts, uncles, sons, and daughters who returned from the cities with their new families. They go in and out of each other's houses freely, carrying messages, household items, or food. Often a visitor does not know where any particular person lives. A grandson will take a shower, change clothes, and eat at a relative's house, even though he lives just across the street. It also seems that from every window, door, or yard, everyone can plainly see what is going on throughout this "village." The African idea of the village caring for and raising children is very much in evidence here. Those children whose parents moved to the nearby cities or faraway places are eager to come to Randolph to be with cousins, grandparents, and other relatives. Many who get sick or die elsewhere return to be buried in the cemetery that used to be segregated. They return for reunions, celebrations, and crises. They never truly left.

As the hot Arizona sun beats down on a newly built playground, the squeals of laughter are beautiful. Little ones taking care of littler ones, all safe and happy and free in their, so far, color-free world. One grandmother sighed, "Lord, they're a burden sometimes, but I have to raise them, can't see them get lost."

A lack of jobs and services and the aging of the early residents from those other states and towns back there in Texas, Kansas, Missouri, Arkansas, Oklahoma—and other western towns such as Yuma, Santa Ana, Sacaton—have brought Randolphians to a place of self-content. The recent encroachment of polluting industries to the east of the town, such as the Sunbelt and Proler refineries, which give off toxic wastes, has pulled blacks, whites, and Mexicans together to fight, if they can, the power structure that is directing one more blow against this community. Many suffer health problems as noxious and toxic materials fill the air. "Homes beneath smoke stacks, schools near factories is not progress," environmentalist Mary Jensen says. "They [the factories] go to towns that have no political clout, the people there don't have a voice, they don't contribute to anybody's campaign. Nobody speaks for them. It's racial and economic discrimination." Studies nationwide reveal that largely minority and low-income areas, like Randolph, are the places where these toxic industries usually set up business. This "dumping on the poor" is endless.[11]

Randolph, Arizona, unlike Boley, Oklahoma, was not established to be different, to be an experiment, to be a municipal monument to the black man's right and ability to rule himself; and so it's not. The war took the young, the railroad took its business elsewhere, the interstate was built to the south and took the traffic, the boll weevil took the cotton, and the machine took the work. A lot has changed since today's residents came to town. Their narratives tell the rest of the story.

Sarah Lucille Smith

Sarah Lucille Smith is the only white resident of Randolph whom I interviewed. I begin the narratives with her words not only because she has lived in Randolph the longest, but also because she is the town's self-appointed historian. Her perspective on the life of this community displays objectivity and truth, and her voice articulates much of what the residents of color could not or would not verbalize. She can be said to embody the "white settler society" that one finds in colonized places such as South Africa, where the whites feel placed yet disconnected. As a part of this community that symbolized another version of the American West, Mrs. Smith never came to grips with or clearly understood the "colored" problem. She is one of a handful of white residents who live on the west, or "white," side of Highway 87. While the "newer" and "younger" whites primarily live in trailers, she has an old yet distinctive Quonset-style home.

I was introduced to her by persons at the Coolidge Historical Society, and our warm and close relationship began on our first meeting in 1991, although I did not begin her interviews until 1992. We did three, and the last one, transcribed

Sarah Lucille Smith's Quonset-style home.

here, was done in August 1994. Between those times, we spoke by phone often, had short "hello, how are you" visits, and exchanged letters, cards, pictures, and gifts. She always exhibited great concern for my welfare as I drove between Randolph and Tucson, advising me about dangers of the road or weather. She always seemed happy to see me, no matter how hot the day or how terrible she felt.

Sarah Lucille was born in Arkansas, moved to California and then to Arizona, where she married Lee Smith in 1947, and became the mother of four children. All were raised in Randolph. Her mother-in-law, May H. Smith, came to Arizona in 1927. May Smith and her son Lee lived in the household of Channing E. Babbitt, the first postmaster of Randolph. May became the second postmaster after Mr. Babbitt's retirement. Sarah Lucille speaks fondly of Mr. Babbitt, who she says was a "good and religious man." Mr. Babbitt supplied wood, ice, and other supplies to businesses and homes in the area. After May retired, Sarah Lucille became the third and last postmaster of Randolph until the post office closed in 1983. She says, "When they closed the post office, it didn't hurt the town. It was already hurt."

On my first visit with Mrs. Smith, she struck me as a bright, well-read, honest, and caring woman, and as our relationship grew, my first impressions of her proved to be correct. A proud part of her biography is that she was selected for inclusion in Who's Who of American Women *while she was doing government work.*

The essence of Sarah Lucille is her connectedness to Randolph. She actually said, "The village does raise you." Also, as postmaster she ministered to everyone and knew a lot about each person in the small town. She was mother, sister, cousin, aunt, nurse, chauffeur, adviser, consoler, confidante, and interpreter for the residents. There were few secrets, and when there were, she kept them. As her words flowed about her neighbors' various experiences with poverty, prejudice, or joy, her body language showed a deep affinity and familiarity with the people and the place. She is a woman of conscience and while, indeed, she lives on Randolph's white side, she knows intimately the life on the black side. Like Eleanor Roosevelt, she exhibits a strong sense of right and wrong. Her empathy for the women and children shows deeply as she shakes her head and says, "They worked so hard and suffered so." At times Mrs. Smith refers to Randolph as "down there," as if it were a separate town or place, and to the black people as "them" ('em) and "they." In spite of her bond with that community, she possessed a naïveté of how the language and the things she said could be received as condescending or racist to a nonwhite person. Yet, I felt that she meant no harm, and this was her way of expressing her honest feelings.

Her face is that of a survivor who seems to have had her share of personal pain. She said, more than once, "I always seem to come up against hurdles, you know."

For five years we maintained a long-distance relationship between Randolph and Missouri. Her letters always contained some information for this project. There would be clippings from newspapers, a picture, news of Randolph, or rumors she had heard, leads to follow. She saw this project as important and believed it would give Randolph a visibility it deserved. She protected any new information she had for me and was leery of strangers who asked too many questions that were intrusive to my project.

In one of her amazing, touching, and funny letters, dated September 22, 1994, she sent me a clipping of a J. C. Penney ad for a Thanksgiving tablecloth. She couldn't wait to share that picture because it had made her mad. At two o'clock in the morning, she wrote that the tablecloth actually had black cotton pickers on it. She wrote: "I am trying hard not to be 'opinionated' in this. There is an old saying, 'If I can't be the tablecloth, I don't want to be the dishrag.' I know if they hadn't gathered the cotton, there wouldn't have been food on the table. I see where this cloth is in the 'Harvest' theme of their products. If you look real close without a magnifying glass, you can see the black faces and straw hats."

I immediately rushed to the nearest J.C. Penney store, only to see that the design of the tablecloth wasn't what she had thought, but consisted instead of fall foliage, pumpkins, and flowers. I told her jokingly that it was the late hour that had fooled her eyes! But the key point is the extent to which this incident indicates Sarah Lucille's social conscience. About the black and Mexican migrant population in Randolph and surrounding communities, she says: "These poor people here didn't have nothin'. I think they were brave to come here. . . . There's no tellin' what was told 'em back there where they came from." She continues: "I look back and I think . . . why didn't people want to help people?"

Mrs. Smith's illnesses began to show more and more in 1994 and 1995, and it was painful to see her in such poor health. Our last correspondence was a letter written from an Apache Junction, Arizona, long-term care facility, dated January 17, 1996. Despite insufficient and erroneous addressing on an unsealed envelope, the letter reached me. It was fated to get to me, and I felt joy and sadness in hearing from her. She said: "I've closed the home; I just couldn't live alone anymore." Sarah Lucille Smith, concerned citizen, patrician, and Randolph historian, died on May 5, 1996. She was seventy-five years old.

In the interview that follows, my occasional questions are given in italics.

There was an old Carter family that lived around here as members of a strange religion that met down there once a year. People came from all over the world to that meeting once a year. They started sendin' things one month ahead of time through the post office—hams, eggs, everything—because they came from all over the world, and they had no building. They do not believe in a building for their religion. They just met out in the open. They called it Truth, T-R-U-T-H. They would live down there, right down at La Palma, south La Palma, just thousands of 'em.

You know, we didn't have water here, at that time. Mr. Jordan was my mail messenger. Oh, for about five years he was a good one. He met the train twice a day. He took mail down on this thing and brought it back up when they dropped it. He hauled water at fifty cents a barrel for everybody. That's how they [the black workers] got their drinking water and everything, honey. He went over here about a mile and a half, and he just had ol' rusty barrels, and that was all he had to haul the water in, and then he would bring it and then put it in their barrels at their house. They just paid him a dollar for two barrels, and you know some of the poor people (this was before, remember, Social Security got around), they didn't even have the dollar to pay for drinkin' water.

And then the Carters drilled this well down here, and Mr. Woods helped 'em by furnishing 'em with groceries. Anyway, the water company then bought from the Carters. But our first electricity was twenty-five watt. Everything you had was twenty-five, the motors, refrigerators, everything. So we were glad when we got hooked to sixty watt.

I believe that the electricity came soon after Lee and I were married, about 1949. So prior to that, we didn't have natural gas. So when we bought this place here, it had a well and the water was piped to twelve families on this summit. And I said . . . well, we went to a lot of expense, and if I ever get out of water, the people will be the first to be cut off.

The Thompsons were quite a close-knit family. There was a father and son and daughter in the beginning and then, of course, they had their families, and they were businesspeople. I know they had the first bar, and at that time the Indians couldn't buy, and these Indians used to make some frequent trips. He had a Chevron station. Now when the Marshall boys came back from the war, they put in their Texaco station right here where you see a radiator shop now. They bought two lots from the same woman that we bought from. She tried to state in it that they would not make it a service station, just a repair shop.

Mr. Babbitt, many years ago, about 1926, and Governor Hunt and a senator had a meeting north of Coolidge, and it was decided then that Florence would get the state prison, Phoenix would get the state hospital, and Randolph would get this old girls' reform school. Now Coolidge hadn't had anything that was put out by the state.

But I remember May Smith tellin' me how these girls never give anybody any trouble down here. They would bring fresh vegetables from the prison over here for 'em to eat, and they would always come and leave Mrs. Smith a big bag of food, groceries. Mrs. Smith is Lee's mother. They were the kind of people that would help just anybody, you know. But I never will forget he was very fond of my son. I had Ray Dunn when I married Lee, and Mr. Babbitt was very fond of him. He went to the track with him and to meet the train. Ray Dunn was ten, almost eleven then.

I gave Mr. Babbitt's teaching certificate to my daughter . . . 1918 one of 'em was dated. It was written on just a plain piece of tablet paper but signed by everybody. Beautiful writing. He was eighty-two or eighty-five years old when he died. You know, from the time Lee and I married, I never knew him to be without this little outfit of Mogen David wine and syrup of Pepsi. He said his first wife's father was a doctor. As you got older your digestive juices change and you lose 'em, and the nearest thing they had come to perfecting those was the Mogen David wine and the syrup of Pepsi. He would take a teaspoon of that and a little ol' outfit of that before he ate. Never had any stomach problems. But when he finally died, there was a tiny hole in his gall bladder. That was the only thing ever wrong with him. He had all the old remedies for different illnesses. He got them handed down. When I showed it to my doctor, he wanted a copy for him and his dad.

The architect of the Arizona Training Center was Lester Mahoney. It used to be called the Children's Colony. The organization that arranged all the money and everything and the building, construction of the children's program, lived with us here for nearly five years. And I do know that every load of cement that went by down this highway to that place was inspected, because he came home at five o'clock and they thought, uh huh, Mr. X has gone home. He won't be checkin'. We'll just bring in a bunch of concrete. If they could bring in poor-quality stuff, they wouldn't have to make sure it was good. He'd sit out on a chair right here and he'd see a truck go and before you could say "scat," he could tell if it was good or not. I believe that was in '51.

The people that work there, at the Arizona Training Center, are endowed with somethin' that I'm not endowed with. I couldn't be a nurse, it would just kill me. I couldn't work around these children because it seemed like I've got to be able to show that I've done something. Accomplished something. With them, like you say, if they just move their hand one way and they did it, well that's great progress.

And the way it was here, every time you wanted to do somethin' out in the community, one of the welfare women would say, "Now this could get political, so you better be careful." I'd just barrel on through.

My husband and I were involved with the icehouse. We sold ice, and Lee left the ice there in a little store that I was in quite a bit and traded there. Seemed like they were Jordans who owned that little store down there. But I will never forget Mrs. Cleveland [Myrtle] Jordan when her father was real ill down on the reservation. I felt so sorry for her and bein' married to the Negro. She could go down and see her father, but she could never stay all night on the reservation. So she'd go down and have to leave at dark. Well, she was Indian, and that was one of their things. She just could not stay since she was married to a Jordan.

The total population hasn't varied very much, just about 250 by now. I would say fifteen to twenty people are white, because just as soon as they grew up and was able, they moved out of here. Just nothin' to do. And that's the same way it is with these kids. These kids here now do not have a place to play or go anywhere for recreation. They've got to get there on the highway and get a ride into town. There's nothin', no jobs for 'em out here. If they get a job, it's a little job in Coolidge or Casa Grande. There's no buses. We are almost flat down to nil. And I told one of the postal supervisors who came. He said, "Well what do you plan to do with the post office building." They didn't want to pay enough rent, see. He wanted me to pay the electric bill and everything, and I just told 'em, no, I was gonna have to put on a new roof and paint and it ceased to be a good investment for me, and I didn't want to be bothered with it anymore. I said, "Randolph is a town that progress just passed by." It just went by us, and the stores just closed up and everything. Of course, as long as we held that little ol' post office here, it really didn't do anything. I look down there and see the people sittin' out in front just doin' nothin'. Nothin' to do. When they closed the post office, it didn't hurt the town. It was already hurt.

It was May Thompson who wrote the society column in the *Coolidge Examiner* in the 1940s. All of 'em were members of this Desert Women's

Club, and they were more on the society level. My friends went to Masonic and Eastern Star, and I guess I was the only Eastern Star in town. The others were Rebecca's friends and all. Of course we had our meetings in Coolidge. There was nothin' to do in Randolph, just a little baseball court down there.

I worked with the welfare women one whole year, and Lee went down and we set the swings in concrete, we fixed a little ol' park. Myrtle Jordan donated the property. This was just before they began to help the Negro women get able to meet the public. They didn't know how to sign up to watch that playground for a couple of hours each day and each mother do somethin'. It would be nobody to take charge. So about that time I said to two welfare women, which I thought very much of and they devoted so many hours here, I said, "Please go in . . ." This is Bernice Bell. She's the granddaughter of the one Bell Street was named for. I said, "Please go in her house and show her . . ." She had seven or eight children at the time. I said, "Just take boxes and put the kids' names on them and show them how to keep their clothes separate." So a girl came in and she spent a week or two weeks doin' nothin' but that, and got the children's clothes all fixed in line, and worked with 'em, and also, another helped 'em with their hair and their body odor. I think it was just the plain speaking about it that bothered 'em. They didn't feel like they were capable. It gave them a little uplift. And by God, one or two got little jobs and one or two got some more and it really did help, so they learned how they could teach themselves, hygienically, and how to keep their houses clean. The kids learned, too.

There was a camp, and I think they would take seven or eight kids from here. They had a summer camp. I forget who sponsored it. It seemed like it was the Kiwanis. But they didn't have toothbrushes. They was supposed to take a toothbrush and a bar of soap and such. They didn't have a toothbrush, they didn't have their bar of soap. So I went down and I got all that for 'em, and it was one of Bernice Bell's boys, her oldest boy, that was goin' to camp. And the morning that they left we was all gathered down there and all his little ol' brothers and sister, they was singin' . . . "Bodie's goin' to camp, Bodie's goin' to camp!" But they was just all excited. Some of the mothers were very protective. I told 'em, I said, "Now about the end of the week I'm going with these welfare women to see how they are doing." This was in 1958 or 1959. I'd say, "I will go with them up there to the mountains and see how they are doing myself." So when I went, the first two little kids came up to me and said, "We want to go home." I said, "Honey, I can't take you home, you got to stay here two more days. They will bring you home. We don't have room in the car." I couldn't tell him that I just couldn't take him for any other reason. He wouldn't have understood.

The first original school was Kenilworth way over here east of Coolidge, which is now a Christian school. Then they integrated. Lee, I think, went to Florence in his high school days and the other kids went to Coolidge, but this highway was sort of a dividing line as to where they went to school. Borree Corner is over here, west of Coolidge, right straight out that road. You know, I don't know why, but this first road north from here, a mile from this one down here, is named Randolph Road. Seemed like this one down here should have been, but this one was named after one of the guys that was a Kleck . . . so that kind of gets everybody off their course when you say Randolph Road, you'd think, well it's somewhere up in here.

Do you know of any other intermarriages here? I know there are ones between Indians and blacks.

The Roof sister married into the Fields family. *The white Roof married a black Fields?* Yes, a daughter married him, and he was a commissioner or city councilman in Coolidge. I think he's also a high school teacher. He went ahead and got an education. Now one of them died not too long ago in the Fields family. I saw it in the paper.

You hear such things that are embarrassing to you, and you know how that person feels inside. In Coolidge, there used to be a group meeting in the little ol' Coffee Cup Café where I ate breakfast, and Wayne Roof always came in and left a tip for the coffee and stuff. The men were talkin' and laughin' about how their luck was changin', and they said there's an old sayin' that if they went with a colored woman, you know, and got with her, it would change their luck.

So when Wayne walked in and sat down to have coffee with them, they said, "Well Wayne, we're all gettin' ready to come out to your place so we can change our luck. Can you fix us up?" And I just thought, oh, my God, poor guy. He never sat down with them again. They just said it before they thought, 'cause they knew about the daughter. They had a beautiful child. Wayne's wife is a teacher too. She went to college and got her degree. I've seen them together quite often.

We had some VISTA workers, two women and a man. Somebody donated a little ol' house down there, and they'd go around and show 'em how to fix their doors, when it fell off the hinges. You know, some of 'em down there didn't even have screws, and they didn't even have a screwdriver. One of the white VISTA girls got pregnant for a black man. I don't know what happened after that.

But the VISTA workers really did good here. They showed 'em how to do things around their home. That was what was important. That someone took interest. And they even started some kind of a little ol' school over back there close to the railroad. Somebody had a building donated. They came out and they taught some of the men how to use these different tools and paid 'em a little bit, but nobody just ever took a great big interest and hired 'em after that, so there was no place for them to go to work.

During this time, I helped get the bathrooms in there, septic tanks, cesspools built through Farmers Home. We worked together on that. One of the VISTA workers brought me all the checks to pay the plumber when they finished. Of course they were government checks. The people seemed to think that they would rather have the money rather than fix up the houses. I told 'em, "You're gonna have to go in here and show them how to keep them clean, what to clean their stools with, how to keep their hair." 'Cause the grease and stuff would stop up their pipes. And they were gonna have to be shown how to utilize what they have. So they did that, and one poor lady . . . she's dead and gone now, she come up there and I felt so sorry for her, she said, "My lavatory just jumped off the wall!" I said, "What happened?" She said, "It just jumped off the wall, Mrs. Smith." Well sure enough, it was cold weather, and it was just hooked up. And ours are connected with a pipe. This, however, was hooked on a clamp, and just like that, with expansion, it came off. So I called Tucson and I said, "Now you better get down here. If they haven't secured these pipes and things a little better than what I have seen and heard, well you better hold up the checks."

My interest was to make people's living habits better. To make living conditions better, cleaner, to give 'em hope. I didn't realize it until later on when I quit on disability.

During this time, the black people here were still picking cotton. They didn't have cars to go to town. They'd have to hire somebody to take 'em into town. I tell you what the white people didn't want . . . these store owners didn't want 'em to get any better, 'cause they'd go into town to buy, see. Keep 'em down. But you'd like to go different places, you know, like to go and get fresh vegetables, and when you face reality, that's really what it was . . . they didn't want them to get any better.

Can you tell me who owned those two nailed-up stores?

The Griffins owned the Randolph Market, and A. E. Woods and his wife, Emma, owned the Woods Grocery, the welding shop, and gas station. The whole Woods family were in business.

My husband and I had the ice franchise for the whole Pinal County. We bought fifty-six brand new vehicles while we were married. We went back to Detroit and picked up several that we finally got rid of and sold them. We had four drivers. But at that time, see, we did not have the ice machines, and the cafés and things were just yellin' for ice. I would get up at three in the morning and go with one of the big trucks to Chandler and bring back a load of ice. Big ol' three hundred blocks. Just couldn't get enough ice in here for the people.

And some of those poor workers. Lee would want me to ride around with him for company. And I'd say, "Lee, it just breaks my heart." They'd be comin' in from the fields about the time he made one camp. Honey, it was like Santa Claus had come. They would be so sweaty and so dirty and so hot. The only thing they had to keep their milk, their eggs, or a little drink of water, was a hole down in the ground. And they boarded it up and that's where they put a lid and a cover on that and locked it. And he had the keys to all of 'em. That's where we would put their ice—in a hole in the ground. He run 'em clear for a week, a month, you know. There was no money exchanged there or nothin' 'til on weekends when he'd collect. They knew that they could get ice if they didn't have a penny. I said, "I just can't understand to see 'em like that." One poor woman looked like she was gonna have her baby just any minute, and the sweat was just pourin' off her.

They would go out and chop cotton, you know. Cotton season starts way back as soon as it comes up out of the ground. If it's too thick, they've got to go through there and thin it. If there's weeds, they've got to catch off the weeds. So it starts from the time it sprouts 'til it's gathered starting in September. They plant usually by February or March when the ground is warm enough. Now they have begun to raise different crops over at Maricopa like potatoes, onions, turnips. Picacho is a big place that a lot of 'em go to work. But then Social Security came in, and it helped people so much, not to have to do that.

Which of the white families owned farms and hired on black, Mexican, and Indian workers?

Mr. Munday. He was the one that owned this farm and this back here. I guess Kleck did too. I didn't know too much about Kleck. He was always on the other side of the tracks, you know. But there are still some with their big farms. There were also some black-owned farms, you know. Mr. O. C. Fields, Cleveland's dad [Jesse Jordan], the Dostys, the Whites, and some other ones. They hired their own people or worked it with the family.

Why was Randolph never incorporated?

It was never even mentioned. When the VISTA workers were here, they were so busy with tryin' to improve the conditions, there wasn't any time for it I guess. After I got out of the post office I believe they tried to incorporate. I wasn't even here at that time. I think that's when I was over at Stanfield. I managed some apartments over there when my little grandson had leukemia. I believe they tried to get it, when they got the street names and everything . . . , they had a meeting, but Twilight Trailer just closed it down. They didn't want to be a part of it. This region and La Palma was gonna be put together.

But this area was not important enough. In fact, there was an article about it ten years ago that came out in the Tucson paper—Melvin [Moore] had it put in. The post office building which was built in 1955 was the only building that had been added in all these years. Because there was no water, no nothin'. Many of the people then lived in the cotton camps. A farmer would have a whole bunch of little cottages, and he usually had one water faucet out there for all of 'em and that sort of thing. But these poor people here didn't have nothin'. I think they were brave to come here, I really do. Not many people would have come here under these conditions. There's no tellin' what was told 'em back there where they came from. Maybe someone in their family had this house, and died, and somehow they come for it. And some of those older ones saved every penny they made out in them fields. They did not spend it until they got money to buy a lot and set up a little outfit. Some of the farmers, instead of supplyin' 'em places at their place now, have gone to supplyin' them with a trailer if they will find a lot. So that's what some of 'em have done now. And anywhere you see Hispanic workers, you will see green stuff. They can grow stuff [laughter]—they grow trees and things. They can use water galore.

In the old days, I don't know why we didn't have more to die from, typhoid or somethin' here in Randolph. Wind would blow those covers off those old dirty barrels. Those Quinteros lived directly across from the Roofs Welding Shop across the highway there, the father had a lung out from TB. The son had lost a lung. A daughter-in-law had TB of the leg. This is just in one family. I never will forget these things. Nobody would do anything to help people. I look back and I think . . . why didn't people want to help people? They would give 'em my phone number if they had anybody in the hospital, or at the TB Sanitarium in Tempe or wantin' to know if it was possible for me to bring his family up. I said, "I'll be right there." And I got on it and picked up the mother and daughter. I mean to tell ya, it was a Sunday afternoon. I didn't get back till midnight that night. But I got 'em up there. He said, "Thanks, Lucy, for bringin' 'em." So they stayed with him about a couple of hours. Do you know when I left he was alright! He was just about gone, too, when we got there.

Then this little Looper boy had cancer of the brain. They had to have somebody ready and available immediately. Soon as that stuff came in on the plane at Sky Harbor, we would leave and get up there to use it. And you have to use it real fast. I was there alone with him, you know. This May Thompson, what did she do? She called our family doctor and asked was there any danger of bein' around me in the post office with all of this stuff that the boy was given. And he told her, said, "No, there's no problem." So you see, some people's mind was workin' and others' was doin' somethin' else.

The Acunas are Mexicans. They are a pretty big family and have been here awhile. They live on the fringes. Manuel Acuna and Frank Acuna, that was the old family. They were farmworkers. In fact, one of 'em still works for this fella that's got this land. There's one Mexican family over there and one down by Myrtle. Down that dirt road, there used to be a refinishing plant that did real old antique furniture, and they sold old antique stuff. It's quite a thing down in there . . . his wife was only forty-two or forty-three. She died all of a sudden. She died with cancer of the lung. One of the Jordan boys tried to prosecute Proler when we was havin' these meetings, but some shushed 'em. When the fumes blow out at night, and you sit out here you can see and smell them burnin' whatever they burn down there.

Did you know of any women over in Randolph who were midwives?

Yes, Clara Bell was one. She is still living. I saw Metris [White] the other day in town, and she said, "I just got back from seeing Clara Bell and she said to tell you hi." I said, "Well that stinker!" She's in Los Angeles. A lot of 'em wound up movin' over there. Some of their people too. Better living conditions, maybe more money than their pensions or somethin'. But anyway, I told her that the next time she came she should bring her to see me.

Mrs. Bromley was also one, after Clara Bell. This one little girl died. You know they had such a strange way of doing things. On the way to Florence, they had a flat, and they didn't have money to fix the flat or didn't have a tire, and they were there right in Coolidge now about one and a half hours before they could put another tire on it. She and the baby died on the way to the hospital. So they began to investigate, and have some meetings on birth control. I remember one doctor from Eloy gettin' up and saying, "I will pay for every suppository, everything, if the women will only use it." These little babies come into the world without even a chance. I remember coming back from a meeting one time with them with the two welfare women, and we had Mrs. Bell and others with us. . . . On the way back from Eloy about a mile out of Eloy, there was a cross burning near the house. You see, you sure did run up against a lot of prejudice. Well, it's a shame that in this day and time people do hurt people like that.

What other Randolph stories or incidents remain vivid for you?

There's been two things in my lifetime that I can never seem to get out of my mind. One was this Mr. Dosty. He was a good man. He told me how when he had smallpox or somethin'—he was runnin' a fever, and his mother absolutely had to work. Him and his brother had it, and they were in bed together. He said they were seven, eight, or nine. He said, "I was so sick. The cats had begun to eat on him." His mother happened to come home at noon, to check on 'em. The cat had got in the house and was beginnin' to eat on him. I said, "Dose, no." He said, "Yes. Mrs. Smith I know it's terrible." He says, "I want you to know it." I said, "Honey, I know what you went through." The other one was just Bernice Bell when she come in there and was holdin' that baby at eight-thirty in the morning. I had just opened the office. She said, "Oh, I've got this cold, too. I've got to get a ride to the doctor." I said, "Well honey, what's wrong." She said, "The baby's guts are coming out." I said, "Did you let it fall, or did it cry, or what?" She said, "No, my other children sucked its navel." Because of hunger I guess. That was two things I just can't forget. I tell you! I know it happens. I had heard of it happenin' in animals, you know. It happens in animals, and that's why sometime they separate 'em real quick or somethin'. Those two things . . . I just sat there like that. Somebody gave her a ride immediately when she got out there. The next person that came in asked me, says, "Lucille, are you sick?" I said, "I guess I am." That's just all I said. I hadn't moved, I hadn't done anything.

I know this young Robbie Dosty, he is a successful black Randolph athlete. He went somewhere and spoke to a group here a couple of weeks ago, maybe a month ago. I just thought, I'm glad to see him back in the swing of things. He was such a great basketball player at the University of Arizona. He spoke to a group of juveniles that was givin' problems. I just thought, I'm so glad to see him back in there workin' for us. Robbie's mother was another case. There was a strange case. She kept goin' to the doctor, goin' to the doctor. I kept seein' these little white spots. I knew that they ordered awfully heavy cream, bleaching cream and stuff. Mrs. Griffin would order double strength and this and that. She had scleroderma. At that time, nobody knew a thing. The library wasn't even a library hardly. So a guy that was sellin' books brought me a book about it, and it's where the body turns to cement. First, she noticed it in her arms here. You would touch 'em and they'd be just hard. And then when it got into her lungs and heart, you know, everything just hardened. I asked the man and the man said, "They don't know what caused it or anything about it, other than it might have been a poor diet when they were growing up." It's where it just turns just like cement. You couldn't penetrate it. Then it got into the breathing cavity. I took her egg custard. She was in bed, and her bed was just as white and clean as could be. Whoever was takin' care of her was doin' a good job. I took her a little bouquet of flowers. She always wanted Robbie to behave and be nice, and polite. Robbie wasn't very old when she died.

One Negro woman in Randolph I remember was quite remarkable—Melvin's mother. She was a worker and contractor. Her name was Beulah. She was a very nice lady. I don't think I ever saw her wearing pants; she wore dresses all the time. Maybe she did out in the fields, I don't know. She was very devoted to seein' that her children got that education. That was the main thing. She worked and sent Melvin to ASU [Arizona State University]. Of course, the others was married and out havin' children, you know. Melvin was the youngest. So he did get his college education. She worked in the field. She had a little ol' pickup. I guess when you see one with a cab on it, you think it would probably hold about six people or eight people. She hired her own people. But she would load that up with her water and food and everything and take those pickers out, day after day after day. You see, the farmer would pay her so much to clean his fields. It was always her and these two boys and the other boy got married or had a baby, or somethin'. There was a brother there. It would have been a brother to her husband.

Alice, my friend, gives me the Hope, Arkansas, paper to read. I see her every once in a while. Wil Muldrew told me that somehow they spelled Muldrow two different ways, you know. They were real nice people, similar to Melvin and Melvin's mother. They were from Hope, Arkansas, where Clinton's from. I had a picture that Alice brought me from Arkansas. The sign going in to Hope, Arkansas, says, "Home of President Bill Clinton." I wish he'd come out here to visit. I don't know how many is here from Hope. All the people originally down here is from the same area. And you know it's odd how I wound up here, because my father was a forest ranger. I was born in the ranger station in Hot Springs, Arkansas, when he was a ranger right there around Hope and I wound up here. [Laughter] You never know why things like these happen. Maybe it was planned for me to be here at this time to help 'em and show 'em that they are still some good ones of us [whites] still around.

Manuel and Eva Acuna

The Acunas are one of only a small number of Mexican families who have resided in Randolph for more than thirty years. The presence of blacks as neighbors did not diminish their comfort with living there. They have been friends with many black families, with whom they have watched and raised each others' children, visited and celebrated together. In Randolph, they have felt safer than in other surrounding communities, and they also feel that their seven children have enjoyed and benefited from being raised in this small, primarily black town.

The Acunas moved from farm to farm because of, as Eva says, "her husband's bad temper and impatience." She, however, wanted to lay down roots and "have a place with water rights and all kinds of animals—pigs, rabbits, chickens—so that the grandkids and great-great-grandkids could touch them." So when they bought their place in the northeast section of Randolph, that was it! Eva continues,

"I didn't want to lose my kids' mail . . . when they wrote to me I wanted to be here to get it right away." This pretty much indicates Eva's strong commitment to family.

During the interview, Manuel was the quieter of the two, interjecting only brief statements to extend or clarify Eva's colorful narrative. Eva was delightful, spirited, and funny, and her speech patterns were very much in keeping with her cultural traditions. As you approach their house, you are greeted by the sign "Los Acunas" and by a joyful, comfortable smile. They wanted to know if I was a visiting Randolph relative, and which family I belonged to. Each time I visited, there was always a son, a daughter, or a grandbaby around. Theirs is a happy home.

When asked about racial or class prejudices they had experienced, they both agreed to not having had direct or obvious encounters with prejudice, but they did recognize that anyone belonging to an ethnic minority cannot miss the prejudiced language—physical and verbal—of the dominant society. She talks, for example, about the woman who kept referring to Martin Luther King and Cesar Chavez as models for "you people." Interestingly, they go on to say, "We're Mexicans, but we don't know what's going on in Mexico, you know. We understand their language, but we don't know their customs. . . . We never go over there. . . . We're not mean to them or anything like that . . . we just don't mess with them." Both Eva and Manuel were born in the United States—she in Santa Ana, California; he in Yuma, Arizona. Some of their children married Mexicans; some married into other races.

Like all the other residents who worked with cotton for many years, Mrs. Acuna picked; Mr. Acuna did irrigating, mostly. They decry the fact that there are fewer migrant workers now than when hundreds of people used to come from Texas to Pinal County to pick cotton. "Now there's nothing, you know. The machines have taken over what fifteen or twenty people did; now one machine does all that." Although Mrs. Acuna did not like picking, she realizes that it produced a good livelihood for her family. It was their life. Like many migrants, they would go to California before and after cotton harvesting to work. In essence, they followed the sun and the crops.

But they are happy in Randolph; it's home. Eva says, "We make menudo, we make tamales at Christmas, and we have a big pan, so when Randolph sees that pan cooking, they know that they can come and get some." The Acunas are very special, and their words validate their love for their family and their community. "We don't regret buying here and staying here. We feel that at least we have a little freedom here, where other people don't have that in other towns. They're so crowded you can't even go outside without stepping on somebody's foot. . . .

We're not related, but we can go to any of the homes here. You knock on the door and they say, 'Come in! Come in!'"

E: We've been here in Randolph about thirty-four years. When we first moved here, there wasn't that many people around. The people that were living here, they passed away already, so just a few like Mrs. Reece are the only ones. In this area right here there were just two Mexican families; the rest were black. The Youtseys, they lived right there, and this other family, the Robles, they bought the place there, and then the father passed away. That was about the only Mexicans that lived around here. But on the other side of the highway there were just white people.

M: One Mexican family lived across the highway, and there was my brother, too. But he moved to Coolidge.

E: We always wondered why we were over here and they were over there, you know. There was a line between us. Nobody ever told us . . . we never went deeper than that. We just thought about it.

M: Right now I think there's only one white family over here on this side.

E: They've been here about one or two years. Now when we first lived here, it looked more like a desert. There was no water. So when we moved here, we just moved and we liked it because nobody messed with us. We never locked the door. Everybody just left us alone. We left everybody alone. Our kids used to play with their kids, and they grew up together. Right here next door we had a family, they were black, and their kids used to come over here. I used to feed 'em, when they were out there picking cotton. One day the little girl comes in and she says, "Eva, my momma says if you can baby-sit for me today." I said, "No, I'm not, I'm tired of all the kids." And she said, "Well, she said you'd been doing it for a week!" They moved to town. I guess some of the kids that left, went to California and came back, and like Steve White, you know. Well, he left and he came back. He said, "Now there's no place like Randolph."

M: All these people that left Randolph, I think they all went to L.A.

E: Yeah, Stockton, all those places—inner cities. Our lives here have been alright. Like I said, nobody would mess with us.

M: And we don't mess with them.

E: It's a quiet town. We like it. A lot of people, they make all kinds of noise. One day I was at the doctor's office and this lady said, "So where do you live?" I said, "Randolph." "Ooh?" she said. And then I said, "Why?" I said, "You know what?" I said, "Where do you live?" She says, "I live right here in Coolidge." I said, "I bet you can't leave your clothes on the line!" In Randolph you can leave it two or three weeks, and when you want it, it's still there. She said, "Yeah, I've heard about it." You know, nothing happens here that will scare us. We know the people. They know us. We take care of their house when they're not there, and they take care of ours.

M: There's not very many places left like that.

E: Because, you know, like in the desert, not even then you can be safe. Like right here in this area where that trailer is, there was this lady, she was real old. Every two years she would go to Texas and visit her family, and she had some chickens. So I would feed them. When we were going on vacation—we used to take off for three months out of the year to go work in California—and I would go over there and I would tell 'em, "Ms. Jerlee, we're leaving." And she'd say, "Well take care of yourself." I said, "Okay. I didn't order no furniture. I didn't give anything away [laughter], so if you see anybody goin' in my house, they're stealing it." But no, the kids used to leave their toys scattered all over the place. When we came back after three months, everything was the same as we had left it.

We always lived around this area. We used to move a lot because my husband was short-tempered. He never put up with a foreman, you know, trying to tell him this or that. So we moved a lot. All this area around here. We used to live on the ranches. We never lived in the towns. This is the closest we lived to a town. We didn't pay no rent, we didn't pay no utilities, nothin', because he used to work as a farm laborer. So we would move around a lot. Finally we ended here. I really didn't like it at first because there was nothing. The lot was empty. There was no water. That was my main thing to dislike. I like a lot of plants and trees. But then the rumor started that they were bringing in the water. They were gonna build a well here in Randolph. Which they did. We had lived here quite a few years before the water came in. Gas and electricity, it was on right away as soon as we moved in, but the water—I don't remember what year it came. So there was nothing really, but it kept us here because we liked Randolph. Manuel worked on the ranches, doing irrigation and tractor work.

M: I used to like to pick cotton. I was a good cotton picker. A thousand pounds a day. By myself! That was thirty dollars a day! She didn't used to like to pick cotton, but she used to do it anyway. We would work on the farms, ride tractors, irrigating, and stuff like that. Now I'm retired. I had to retire. I turned my back irrigatin'. There were green hoses, you know, and real heavy, the ones we used to use at that ranch. That's when I fell and hurt my back.

E: Some of the farmers are going to start pickin' cotton next week [late August]. It looks green now, but they burn it. They throw some chemicals to spray it.

M: They spray it to dry out the leaves, you know, so they can harvest earlier.

September is usually when everybody starts to harvest. In September you want to start harvesting cotton, but right now there's farmers starting to harvest this coming Monday, tomorrow. Like my nephews, they gonna start tomorrow. They spray it with some kind of water mixed with salt or something and dry out the leaves.

E: Then there were two kinds of cotton. There was short cotton, which grew sort of small, and then it was alright and then the weather would dry it, the cold and everything. Now, they do it by plane. Then there was the long cotton. It was way over there. You have to bend the thing to pick it, and then it was real tiny, like pima cotton. Now you see pima cotton and it's almost like the short. They don't do it anymore like that.

M: They mix it too, so much, with some other kind of cotton, with chemicals, whatever.

E: It used to be that you bought clothes that was made out of pima cotton and it cost you more than the other, 'cause it was so fine. I didn't like it. Short, they used to call it, but I don't know about the other kinds.

M: Well, they got so many kinds of cotton right now, you don't even know the names.

E: Some cotton plants have white flowers. The plants with the yellow flowers is the pima.

M: Then there's that one with a spot on the inside.

E: That's pima. But it doesn't grow that fast. You couldn't find each other, you couldn't even see it, that thing was so big. Then the machines took over.

Were you born in this country?

M: I was born here. I was born in Yuma. See, I was just born over there, and then they brought me around here. She [Eva] was born in California. I'm sixty-six years old. I retired when I was sixty-two. I didn't want to.

E: I was born in California, in Santa Ana, about a mile from Disneyland. I wish I would have stayed . . . I was about ten when we came here. We've known each other since we were kids.

Did you always work on the land, in California and Yuma?

M: Well, in California when we used to go over there we used to harvest fruit, like prunes and stuff like that.

E: We were migrant workers. And I didn't like it.

M: You had to be on your knees eight or nine hours. Well, that's hard work. You're on your knees all day.

E: It's like we were asking for forgiveness. [Laughter] We got it, too, 'cause we used to put in eight hours of work or even longer to make a buck there. What I didn't like about it was the moving. I didn't like to move. My goal was for us to buy a place where we had water rights. We had all kinds of animals. I wanted to have a lot of things for my grandkids . . . some of 'em don't never touch a little rabbit or, you know, pig or whatever, and that's what I wanted. I wanted to have all those kind of animals so they could enjoy it, because we have twenty-six grandkids and sixteen great-great. I had seven children. I had twins. My son Gary is a twin. The other one died when he was three months old. We had two sons that went to Vietnam. Not too long ago [1991] I had a grandson that went to Desert Storm. One of my sons, his son is the one who went to Desert Storm. I had two brothers who went fighting over there in the Second World War. We have been a military family. All that time nothing happened to none of 'em. They came back. They didn't come wounded, or nothing. A little more crazy, but otherwise the same. And now, we don't have any animals. We used to have all that and then we starting getting rid of 'em.

M: There was some kind of an animal that eats hay and stuff like that. That was real expensive. I used to buy calves, and raise calves over here in the back. I had 'em boarded back there.

E: The thing is that, like us, since we were leaving every three months out of the year, we couldn't afford to have animals because nobody wants to take care of somebody else's animals. So now we just let it go.

M: So when we were living here we were still doing farmwork. We would leave the house, lock it up, and go to work somewhere else, like up there in Hillsburg, California.

E: We would pick prunes there. Well, they're plums. They dry 'em. Yeah, they would put 'em through some water with something and then they would just lay 'em out in the sun. Then he got a job here with Worth Bartlet, on a ranch, so he worked almost twenty years with him. Yeah, that's when we stopped going to California. It's when our kids started growing up and they didn't want to go. They all had girlfriends and all that, so they didn't want to leave, because they were working, and then, my two sons took off. My problem was that I didn't want to lose their mail. When they wrote to me, I wanted to be here to get it right away, 'cause we had a post office here in Randolph in the early days.

We had a big family, too, but not all ours. We raised two nieces and two nephews. Then, all in all, we raised fifteen kids with what he worked, 'cause we didn't get no welfare, we didn't get anything for them. My relatives were raising some grandkids. When they passed away, they passed them on to us.

With all those kids, he had to work. We had a day-care center here and I used to work there. I used to take care of the babies for working mothers. Then these two other teachers, they used to take care of the smaller ones and then the little kids, they were preparing them for a Head Start and kindergarten. When Mr. White came and asked me if I wanted to work, I told him [Manuel]. He said, "No." He didn't want me to work. So Mr. White said, "It's only for six months." They train you and then for six months you work and then they bring in another, and they train 'em too. So I told him, "Well, why can't I work six months?" So finally, he let me go work. We were doin' pretty good, but then they decided, why spend all that money and then change teachers, you know, when they had spent so much money on us. All those workshops and all that we did. So they decided to leave us there. So I worked five years.

Were the mothers of those children working?

E: Yes, they were working here at Arizona . . . they used to call it the Colony then, now they call it the Arizona Training Center. They changed the name. Colony looked too negative. So they changed it to Arizona Training School. So, I wanted to work there too as my children started growing up and everything started getting too quiet. The older people passed away, the ones that I used to know best, except Mrs. Reece, and then she's always sick. She's always in the hospital. But we go visit her at the hospital when she's over there.

M: On New Year's, this house would be full of colored people, colored kids, the ones we used to know from little kids.

E: They come and spend the night here with us. Yeah. They come to the party that we have here. They start drinking a lot. I go outside and I say, "I don't want you guys drinking. I don't want you guys to get hurt." They answer, "Mrs. Acuna, we only live right over there." We got used to this place. I don't think we'll ever leave.

I remember Mrs. Clara Bell. She was the midwife that saw me when I had my second child. She was nice. She was mostly alone, you know, but her work kept her busy, because all the ladies here in Randolph used to use her. We used to live out there by Coolidge, and I came and talked to her and she went to the house when I was gonna have my second kid, because we didn't have no money to pay the doctor or hospital. My son said, "Well how come I had to be born at home and the rest of my brothers and sisters were born at the hospital?"

She used to charge thirty-five dollars. She went up to fifty. She got more modern. I used to go visit her later on because we knew her. Melvin Moore and Mandy's mother [Beulah], they used to live right here, but the lady that they bought the land from would never want to give 'em the title to the land. She said, "I don't bother you, you don't bother me." But they wanted the title. Then the government came in to put in bathrooms, because we had 'em outside. They used to scare the hell out of me too. I used to go in 'em with fly spray, whatever, you know. Finally the government came in and you had to show your title to the land and all that before they started digging for the cesspool and everything.

M: They put in showers and bathrooms, and stuff.

E: So it was an improvement. So everybody got a bathroom put in, and that was pretty good. But Melvin Moore's mom, I guess she was sort of the first person that came here because she had a lot of lots here. She used to be a contractor for the cotton pickers and all that.

M: Like Mr. Jordan over there. He used to be a big contractor too.

E: Yeah. They call him "The Preacher." One year we worked for him. He was pretty good to work with. Mrs. Jordan was pregnant with I think it was her last baby. She would take lunch to him where he was working. And since she didn't work, she knew how to drive, so he had it good. Her sons started leaving for California, and they all started moving back too. Ronnie was the one that stayed away, the oldest one, he was the one that stayed more out there, and then his wife and him separated and he came back with his daughter and moved to Casa Grande. Then he started being an Indian, not a black guy. 'Cause he was getting more benefits, he said. One time he came over here to visit us and I said, "Well, so what happened, Ronnie—you didn't like it over there in California?" He said, "Mrs. Acuna, you don't know what it is in California. Every step you make, you gotta pay for it." And so he says, "Once I got back here, my mom talked me into stayin'. We always left our Indian blood on the side," he said. And he started out just black, you know. "So now I know I'm part Indian and part black, so now I'm getting the benefits that the Indians are getting."

Ol' preacher didn't even look like a black guy. He looked more like, I don't know, like if he was mixed with not all black, you know. His skin wasn't that black, and then the way he talked. When he was talking to us, he talked one way. When he was talking to blacks, he was talking the other way. He was alright. He didn't push people.

M: He knew how to handle people.

E: We like it here now. I wouldn't move for anything. It's peaceful. A lot of people are trying to buy lots here. We got a letter from San Diego one day trying to buy up the property. I told my husband we're not selling to no Japanese. They are taking over as it is.

M: When we first moved here, we paid $250 for the lot right here.

E: For all three lots.

M: The old man, well, that was Sarah Smith's husband, Lee, said, "Why don't you give me fifty dollars down and twenty-five a month." I thought it was too much money. So finally, me and a brother-in-law of mine got together and just paid cash for the property. After we bought it, I kept on workin' on the farms. Then I got tired of moving.

E: I used to give him a lot of mouth when we had to move.

M: If you were to quit on the farm, you know, quit your job, and you were leaving the farm, the boss would say, "You what?" He'd want a lot of explanations. Or they'd say, "Take everything and don't come back." I got tired of that. If you was gonna quit on that farm or go do something else, you had to move out of the house, live at an auntie's home, somebody else would move in, you know, and work with the man. That's the way it worked. I got kinda tired of jumpin' here and there.

E: There were times when we moved three times in a month. Well, I didn't unpack. What I did was, I used to give this man . . . if I was to write a book on all the things that I called this man, it could fill many pages because I was tired. One time we moved to a ranch over there, and the boss told me to clean the yard, and then the kids, you know, they would take off to school, and then when they came back in the evening, I would have the pile of all the trash and they would throw it around. So the guy saw me cleaning the yard so he brought some of his workers to clean another part of the yard where there was broken glass and wire and all that. So he told him [Manuel], "Tell your wife if she wants to paint the house on the inside, just to tell me and I will bring you the paint." So he asked me, he said, "Do you want to paint the house?" I said, "Yeah."

So he brought the paint and I painted the house and I made new curtains and all that. About a week later after I did all that work, he comes in. He was irrigating, and the foreman wanted him to go chop weeds so when the next irrigating time came around then he would be ready for it. So he said, "I wasn't hired to work in the weeds—cutting weeds." So he said, "Nobody's gonna make me cut. I'm not gonna chop weeds, but I'll go talk to the boss."

So I didn't wait for nothing. I just started taking the curtains off. I made a thing for it, to make it into a quilt. I put a blanket in it, I made a covering for the quilt and all that. I knew that the next place we moved, I didn't know if the house was gonna have windows or big windows or small windows, I wasn't taking chances. I just sewed it together and made a covering for a quilt. I took all my mom's things. She had a lot of household things. I packed them and everything. I put some sheets over the windows, and about two days later he quit. He said, "We're moving." I said, "Okay." I didn't wait for him to say pack or anything. I just went ahead and packed. I was tired of moving too.

This area right here—the house—we just had the trailer, and we had these two rooms right here and the kitchen. And then one year his family was getting ready to go to California, and I told him, "We're not going this year." He said, "Why?" I said, "You're making good money." He was making $900 a week chopping weeds with the kids. I told him, "Why leave this money." He said, "It's not gonna last that long." I said, "Yeah, but look over there. We don't have no guarantee that we're going to make money." The fruit trees had to be loaded. Then if you did not pick the fruit, the guy would give you so many days to pick it. If you didn't pick the fruit then, he would bring in another family to help you out.

So then the guy would not pay you till you finished the season and everything was picked from the ground. By that time, we were borrowing money from them to buy groceries. So by the time we got our check it was almost gone, because it was almost three months that we were, you know, pulling from that check. So when we got here, we paid our utilities because they were behind, and we bought the kids some clothes and we bought groceries, and then we were broke. So that was my argument there. I told him, "This year we are not going."

And then I also told him, "We're not leaving until you make this house bigger." He says, "So how much bigger do you want it?" I said, "Well, I want my own room there, and then I want the living room here, dining area, and then the kitchen, I want you to sort of like fix it a little bit." So then he started, him and my brother, they started building. When we took off, he had put the roof but he hadn't put nothing in there. So we took off. When we came back, I was saving money, you know, because I was looking for a house to be standing up when we came back. Okay, so we fixed the house and then the kids started growing up, they needed more room and all that, but then they started getting married too young, too.

M: All those children come over here, they keep coming back. But that's how we moved here, and then we bought two trailers—well this farmer bought 'em for me, twenty-five dollars each, and we just set one right here and one over here and we moved in, and that was it. That was it.

E: No more moving around. I had to carry the water, but that was no problem. Everybody had their own water tank and stuff.

M: When we first started messin' around here in Randolph, we were payin' rent up there. We were payin' ten dollars a month, and I thought I was payin' too much. So I said, "No. Forget it." And we moved out. I didn't like to pay rent. There was no use payin' rent.

This was in the 1940s, 1948 and on. I had a bunch of kids I had to support and take care of. I didn't like payin' that rent. We used to have two stores here in Randolph, two gas stations. Now we don't have a store.

E: Leon Thompson owned the store. The one across the highway. He had what looked like a little bar. He used to sell beer and whiskey. Had a big sign there, but it's fallen down now. Right here used to be Ernie and Manuel Marshall.

M: E. O. Thompson owned that gas station. Then on this side there was a Mr. Woods's store right there, and the other one was Griffin.

E: Mr. Thompson and Mr. Griffin's wife used to be brother and sister. We had a post office. We didn't have to go far. Now they charge us, what? Thirty-five dollars a year for the box in Coolidge. And we have to go over there and pick it up. Over here we only paid like five dollars for a whole year.

M: You can get it here but we didn't want to do that 'cause sometime we take off, you know, for a week or ten days or whatever, and when we come back, go over there with a key and open the box. But before, she [Mrs. Smith] used to work for the post office. We didn't have to spend money. Or if you need something for the house, there was two of them, so you had your choice. But right now we don't have another post office.

E: La Palma has the closest store. Melvin had a little store, open about three, four years, and then one day I went over there and Wesley, his son, who was working there, said, "I think we're going to close the store, Mrs. Acuna." I said, "Why? You're eating up the profits! Your dad made a mistake when he hired you." It was too expensive. And then he [Melvin] wanted to retire from police work.

There's hardly any noise here either. When we used to have a lot of noise was when these people right here had a lot of wetbacks and they would bring 'em to work and they would fight. And one day this person comes in and knocks on the door and my son heard him and he went to the door and he opened it and he talked to the guy and then knocks on the bedroom door and he said, "Mom! Mom! Get up!" I said, "What happened!" He said, "There's a guy here who's been stabbed." I said, "What?" And he say, "Yeah. At least he looks like he's real sick."

M: They got him with an ice pick right here, and one time right here in the belly button.

E: That's the only problem we had because the people here in Randolph, we trust 'em, but we didn't know those people. They were coming from Mexico, working for a Mexican contractor. We didn't know what their intentions were or what. And then after they got drunk, you know, they got all whoopy. But after the season was over they went back to Yuma.

We're Mexicans, but we don't know what's goin' on in Mexico, you know. We understand their language, but we don't know, like, their customs, or whatever. So when they come to this side, we never go over there.

E: We're not mean to them or anything like that, you know, but we just don't mess with 'em. I mean, if they need help, like there was one guy said he needed a lunch and water and we gave it to him.

Did your parents come from Mexico?

E: Mine didn't.

M: My mother did, but I don't know where.

E: My mom was born there by Tucson . . . Tubac. My dad was born in Santa Barbara, California.

Where did your kids go to school?

M: Borree Corner used to be a school for blacks.

E: Ours always went to Kenilworth.

M: When they shut that school down they mixed the black with the white.

E: But there was no problem there. Like they're having in school now. Here in Kenilworth they go all together because the kids here from Randolph, they were good athletes. All the schools wanted them. There were some farmers that didn't want their kids to go to Kenilworth because there were a lot of Mexicans and blacks. They were sending their kids here to Coolidge. So when they started closing Kenilworth, they put up a fight because they didn't want their kids to mix. All the kids here were gonna go to Coolidge, but the principal said that he didn't care, that he was gonna accept Randolph kids to go to school because they were good athletes.

M: Oh, if they're athletes, these are okay!

E: We have a granddaughter that is black, and she's real cute. My grandson, the one that went to Desert Storm, he got together with a black girl . . .

M: I got a nephew, he doesn't like blacks. I tell him, well, we got together, you know, with people from Randolph at the house. One grandson is part Mexican and part white. My boy married a white girl. So we love everybody and have no dislikes because of race.

Did you have any experiences with prejudice as Mexicans, in working on farms or living here or in Coolidge?

E: Well, one day we went to Wendy's in Casa Grande and these young girls were there. There were some Mexicans and three white girls eating. Manuel went and found a table and I was ordering. As soon as we came in they started laughing. I passed by there and they, you know, they kept on laughing and looking to where we were. So I went over there and I told 'em, "Okay, so you're laughing at me? Some day remember this because this gonna be famous words

I'm sayin'. One day you're gonna look like me. You're gonna be bald, you're gonna have gray hair unless you dye it for the rest of your life." And I said, "But I used to look at old people like that and I never thought I would come to this stage. So you guys are looking at me now but ten years from now, you'll remember this day and I hope you do." They got up and left. They weren't that young. They were, you know, in their twenties, thirties.

We had like a club. This nurse, when I used to work there in the day-care center, would come and peek at the babies. She used to tell the black ladies, "You guys should be real proud of King. He showed you what black is. And you," she told me, "should be real happy about Cesar Chavez, he is teaching you Spanish, the Mexican culture." "He's going through hell," she said. "You should be proud of that man. Like the black people are about King." She said, "These two men, they're great men, but we don't see them that way. We just see 'em as human beings and all that." She said, "And what happens is with people like them, like Cesar Chavez and Martin Luther King," she says, "people ought to not look at them. They are always trying to find something wrong with them. To criticize them." Because Cesar Chavez—that's the reason he died. Because he used to do that thing about not eating and all that. So the last time he stayed [on the fast] too long.

She used to tell us, "You always look up to Cesar Chavez, and you people look to Martin Luther King because he did a lot for you people to show you what you really are." I thought a lot of Chavez, but my husband and I don't see the same things. He used to come to Eloy because in Eloy they plant a lot of vegetables. And he used to come to Chandler because they have a lot of grapes there. He wanted people to go over there and talk to him, but he [Manuel] never wanted to go there. I wanted to go there. But he says I'm nosy, that's why.

He was here in Eloy at a dance once time, well not really a dance, but people put it there for collecting money. People would say that he was leading the migrant workers. He was fighting for 'em, you know, going through a lot . . . he needed money to keep on with his campaign. But people didn't see it that way. In Chandler they kicked him out of there. They told him they were making good money.

He would try to get more money for the workers. More living spaces and all that because people would go there and they live out in the hot sun and whatever and they wanted farmers to make little cabins, or whatever, so they can live, but people didn't see it that way. So when he died, my two sons and a friend went to his funeral in California.

When he died, it was a heavy thing for his wife and kids. So his son took over, but it's not gonna be the same. And there are fewer migrant workers now, 'cause a lot of people used to come from Texas to here, to Pinal County to pick cotton. Now there's nothing, you know. The machines have taken over what fifteen or twenty people did; now one machine does all that.

M: The machines don't chop the weeds out. They got to have people to do that.

E: But the machines are killing the weeds. So on the farm, they're bringing in more people to chop weeds. Like they don't chop cotton no more but they do chop weeds. But now that chemical that they put in there, that chemical is killing the weeds so there isn't that much weedin'.

M: They spray the ground before they plant the cotton, and that takes care of the weeds.

E: Like us, we used to burn the grass, and we used to carry a five-gallon tank behind our back, and we would pump it and then that thing would spray the Johnsongrass and we had to sort of wash it so that that thing would go down to the root and kill it. It was hard work because when we were working, it was like 120 degrees and that thing on the back was killing us. To kill the weeds. And when we empty that tank, we would go back and fill it up again and then start where we left off. It was hard. We started out with those spray bottles. All day we would be spraying, and my finger would get real swollen and black and all that. Like bruised. One day I told the boss, "Why don't you start doing something about this thing?"

What major changes have you noticed in Randolph?

M: As far as other changes around Randolph, new businesses and all, there was another company that was supposed to come from California. Another plant over there by the Jordans, but I don't know what happened.

E: One time there was a fire. It was a big flame. So I came in, Danny and Manuel were in here, I told them, "There's a big ol' fire out there at the plant." And Danny, right away we went outside, and said, "Yeah it looks like that—it is on fire!" So he called over there and he talked to the manager, he said, "I live here in Randolph and I'm looking at a big flame, so what's happening?" And he said, "Steam." And then I told that guy at the meeting, I said, "As long as I've lived in this world don't you think I know the difference between steam and a flame?" I said, "Those people are trying to con us because I saw the fire, I saw the flame." And then we called over there—"Steam." Steam does come out, and in the mornings when everything is calm, right there at the Arizona Training School, you can see the fog from the thing right there. It settles on the trees. This thing right here was killing us.

M: Yeah, they were going to recycle oil. They said there was not gonna be no pollution like this right here.

E: And I told those people, "Okay so . . . like that flame that I saw that you call it steam," I said. "Why don't you guys put something that will let us know that there's danger there so we can get the family out in time, because if that thing goes up in smoke," I said, "Randolph goes with it, too." But you know, they just said it was steam and nobody could get 'em out of that.

M: There was a guy that was gonna open that business right there on Highway 87, was gonna sell some carpet, he got a license to operate, but they told him no.

E: Yeah, this other lady came around here taking names. She had like a petition. She had to have so many names so she could open a shop. The thing she was gonna have was, she said, "Arts and crafts." She was gonna have some paintings and all that. The reason she liked it right here, she said, was because it was close to the highway and a lot of people pass through there. It never came through. And then the carpet thing flopped too. And we got that letter from San Diego asking us if we wanted to sell our property.

M: But people are funny. Somebody wants to put something like that, you know, a place to sell sodas and candy and stuff. Probably pass by that place. When Melvin Moore used to have that little store over there, everybody started going there, you know, pretty good, but that was including myself. Then I kept on going, but everybody started going a different way. So he had to give it up too.

E: Well, we also thought the town would start developing again like before.

M: Yeah, we thought it was gonna be alright for Randolph . . .

E: People would start coming in and settling here and all that, because I've seen other small towns, they start putting out trailers and people move in, and you know it's a town pretty soon. I was telling him one day, maybe we can clean it. We can plant first, we can start out cleaning, but nobody wanted to clean somebody else's yard. They said, "Let everybody clean their own yards." We were trying to get the trash truck to come and pick up the stuff, but this guy tells me, "You know what, Mrs. Acuna," he says. "Right now you're doing okay. You're throwing your own trash. If you guys get with Coolidge and they send the truck over here, they're gonna start charging you for trash, they're gonna find some way to bring up your taxes." He said, "Before you know it, you're gonna be paying as much as Coolidge." 'Cause Coolidge pays sewer, trash, and all that. So he says, "Just leave everything like it is." 'Cause they started charging to throw the trash in Florence. We have to drive ten miles to throw the trash, and when you get there, it depends on what you got in your trash. If you got tires, they charge you a dollar for each tire. Like us, we only threw some stuff here that we couldn't burn, so then we found this farmer that my nephew lives by, and he had some cattle and they make those big trenches, you know. So he wanted that filled up so he is letting us throw the trash there. It's not only free, it's closer.

Well, do you think you will live here forever?

E: Yes, uh huh. We're staying. We don't regret buying here and staying here. We feel that at least we have a little freedom here, where other people don't have that in other towns. They're getting so crowded you can't even go outside without stepping on somebody's foot. Here you can go outside and do whatever. The best part is we can have parties and nobody says anything because we invite the neighbors. They're welcome, and we just welcome them out here, but otherwise they won't come around. They will stop and talk to us right there. Like Pearl's kids—I was making tortillas outside and they stopped one time and the oldest, he says, "Mrs. Acuna, will you give me a tortilla?" I said, "Yeah, come on in." So I gave him one, and he says, "Do you have any butter?" I said, "No. I gave ya a tortilla, okay!" He comes in and gets in the refrigerator, he rubs the butter on it, and takes off. I said, "Aren't you gonna take some for your friends?" There were about three or four guys. He says, "No. I'm the one that got it. I'm the only one that's gonna get it." Because they were here all the time, you know. They'd talk to us like they're talking to part of their family. They grew up here with our kids.

M: We're not related, but we can go to any of the homes here. You knock on the door and they say, "Come in! Come in!"

E: Yeah, you walk in. There's this guy, we've known him since he was a little kid. His mom died when he was young, and there were two brothers and a sister. We make menudo, we make tamales at Christmas, and we have a big pan, so when Randolph sees that pan cooking, they know that they can come and get some. So he comes in and he brings his little pot and he says, "I know you guys are making something, either tamales or menudo." So he takes some home. The next morning he got a hangover, he comes over here. One day we were laughing because I heard some voices at Christmas, no it was New Year's. I heard some voices, I told him, "Somebody's outside." Manuel said, "I don't think they want to come in." Then I look out the window and there were about three or four of those kids that grew up here, the boys, they lighted the fire to warm the menudo. They said, "See, we're good neighbors." And then my son gets out there and he says, "Yeah, because you want to take some home for your hangover." So it's alright.

We don't regret moving here. We even encouraged our nephew to move here. He used to live out there by Coolidge. His dad didn't want to sell him the property, so we told him, "Buy that one right there!" But there were some white people living there. They didn't mess with us or anything. There were those people, like hillbillies, they spit to one side and they cutting the thing with their knife like that. They didn't bother us.

Now, out there on the Kleck Road there are three Mexican families, three over there, and there's one here, and then there's two right here. And then us. Then the Youtseys. And then this one right here. Then the ones on the other side. Across the street right there.

M: Yes, we're happy here, we're goin' no place!

Myrtle Jordan

Myrtle Jordan with seven of her sons.

Myrtle Jordan, a Pima Indian woman, has lived in Randolph since the late 1940s, when she married Cleveland Jordan, a Negro migrant worker and labor contractor. Mr. Jordan had come from Hope, Arkansas, after his father, Jesse, came first and established a home in Arizona. Cleveland had eight children and a wife before he met Myrtle, and Myrtle, too, had been married, to a Pima Indian with whom she had two children. Myrtle is part Pima and part white, and Mr. Jordan was a mulatto. Their story is one of true love, bonding, the bearing of twelve more children, and a closeness and family solidarity that still exists, after Mr. Jordan's death in 1965. She is the "mother" of fourteen children, eight stepchildren, forty-five grandchildren, and fourteen great-grandchildren (as of 1992).

Because Mr. Jordan was a labor contractor, landowner, businessman, and boss who recruited cotton pickers from Arkansas, as well as from the nearby reservations, I saw the family's story as important for the story of cotton. The

Cleveland and Myrtle Jordan.

Jordan pickers and thousands of others had done hard work but had not been documented sufficiently as contributing to the Arizona and western economies. I also wanted to know the women's stories within this industry and way of life. I had heard about the many intermarriages between Indians and Negroes, so the story of Myrtle and Cleveland Jordan is representative historically. Her story is one of survival, difference, assimilation, and the wealth that comes from children, grandchildren, and love.

Myrtle Jordan's story is complex and arresting because it touches on so many areas: race, culture, ethnicity, heritage, language, geography, and gender. She and I talked on several occasions, but the major part of this interview was done on one of the hottest days of July in Randolph in 1992. I believe it was 117 degrees! Yet she invited me to sit outside under a palm tree to do the interview. Everyone knows palm trees do not give shade! She said, "Oh, it's not bad out here, we'll be okay!" She simply gestured at me as if I were a spoiled, pampered woman with little experience with what heat is or with dealing with the small discomforts of life.

We sat there for two and one half hours! Several grandchildren and a son became concerned, and finally brought us lemonade and persuaded her to take me inside. My recovery was slow. It was a great session as she poured out her story, and I became absorbed with this warm and giving woman. The things that struck me immediately were her sense of humor, her animated way of talking, and her attention to detail. Her favorite exclamation was "Oooowe," which at different times in the narrative meant different things. Often it signaled her "reliving" of a moment long gone. Mrs. Jordan's connectedness to her husband's memory, the black culture, and her own life is what ultimately defines her; she refuses to move from their original house or land because she is one with it and Randolph. I also consider her one of "Sojourner's Daughters," because she lived within the black American culture for so long and loved her husband, Cleveland, so much that she felt that she, too, was "black," something she states in her narrative resolutely, without negating her Indian heritage.

Myrtle Dorothy Williams Jordan, Pima Indian, born July 27, 1921, in Sacaton, Arizona, on the Gila River Reservation, died on Sunday, January 5, 1997. Her wishes were to return to the reservation for her memorial and burial among the Gila Indian community in Blackwater, where she was raised. More than five hundred relatives and friends of every color, race, and class attended the services offered in the Pima language as well as in English. Music was by the choir of the Pima Indian Church, where she was a member, a black Baptist women's group, and an Indian woman folksinger. All celebrated her life. I was asked by the family to offer the eulogy and was stunned by this request. What an honor. I had somehow become an "honorary Jordan daughter," a position I shall always cherish.

I was about seven years old when I left the reservation for good to go away to school, and I went to Tucson to a missionary school. They used to call it "escuela." It's torn down now, but it's down by the river. A lot of Indians went there. Pimas, Papagos, mostly. I stayed two years there. When I got sick and they sent me to Sacaton, and I stayed 'til I was in the third or fourth grade. My health got bad, and then they sent me to Phoenix. I had to be hospitalized, but I still went to school. From there they sent me to the Phoenix Indian school. That's where I completed my schooling.

Well, I never stayed on the reservation. Just during the summertime. That's all. I mostly went to boarding schools and during the summer, usually I'd do domestic work or somethin'. I'd come home for a visit, but that's all. After high school, I went out and worked at Williams Air Force Base for several years. Then I did do a little hospital work. I've worked in hospitals as a nurse's aide. And after that, well . . . I fell in love, I guess. [Laughter]

I got married and had two kids and my marriage didn't work out, so I just left and went back to workin' again. My parents took care of my two boys. I was afraid to stay and I might have more children. I didn't want to put up with it, so I left. So anyhow . . . now he's gone; he died not too many years ago. From then on, I worked and worked until about 1944. Then, I came home, and I don't know what made me come home for a visit, but I met this man, Mr. Jordan. I'd go back to work at Williams Air Force Base, and he'd come over, but he couldn't go in to the base, so I had to come out to meet him. We had to have a pass to do this.

Indians picked cotton. Oh, yeah, they picked cotton. On the Reservation they raised cotton. Some of them did. But I always wondered why they had to go out and pick, you know. I never could understand that. The way I met my husband really was through my dad. My dad was pickin'. He went to pickin' cotton for him. And so, when I was workin' at the base and I came home that Friday night, and I told Dad, "You goin' to field tomorrow?" And he said, "Yeah, I'm goin' to field, but I'll bring you back at noon. And then later in the evenin' then we'll go to town and get the groceries." So I says, "Well, Daddy, can I go with you?" And he says, "Sure, what you gonna do? You never picked cotton before." I said, "I'll just see how you all do it. I just want to try." Anyhow, I went out there and that's when I first saw the man. I said, "Hm, nice lookin' black man." [Laughter] I guess he said the same thing about me. He look and he say, "Um, good lookin' Indian." I was in my twenties, and I guess he was in his twenties too. I was out there tryin' to pick cotton and didn't have but that much in my sack, and he says, "You must have never picked cotton." I said, "No, I never have." And he asked me, "Where do you live?" I said, "I work, I don't stay out here on the reservation." Anyhow, he said, "Is that your dad?" I said, "Yeah, that's my dad."

My father considered himself Indian. His mother was Indian, his father was white. And I speak some Pima. Some things I don't know. I have to kind of mix it up. I can talk to 'em, but like I say, "where you goin' " and "what are you doin'." I remember my grandmother. She used to have a big basket on her head. And she'd have what we call *pinole* in Spanish. It's a wheat drink. You either mix it with milk or with water and add a little brown sugar to it. And she used to say, "hiji, hiji, hiji." And with Ron and with me she'd be way out walkin', we'd say, "hiji, hiji, hiji." We didn't know what that meant. We just heard her sayin' it. So we'd be "hiji" all the time, and it means move back, slow down. And I didn't know that. We just called each other "hiji." [Laughter]

If I ask you what is your name, you'd say, "Chat chugig." And like, where you goin', "Pap te hi." We didn't know Pima until I guess I was about twelve years old before I kinda picked it up. I knew a few words, but I couldn't put it together. I said too much of that "cho cho cho" and all that. [Laughter] And then at school some of the girls would say, "How come you don't talk Pima." I'd answer, "Yes I do." And when I'd try to say somethin', they'd laugh at me. They said, "Well, you got to learn."

Every time I'd come home to the reservation, I couldn't stand it, no bathroom, no place to take a shower; I couldn't get used to it. You know, gettin' in the tub or goin' to the canal or somethin'. But I've been to the canal swimmin' with the other kids, and we used to go the river. But I didn't spend too much time there. My great-grandmother, I remember seeing her in long skirts and gathered skirts and blouses, but we didn't dress like that in my time. The government would help the Indian people, you know. That's the way I got a job on the base.

Finally me and Cleveland just got together. And we've been together since 1944. But you know, we could not get married until 1950 because they didn't allow Indians and blacks to marry. No, no. It was Arizona law. He wanted to take me back East, and I refused. I was scared because they told me that they'd do somethin' to me or kill me. Oooowe, a lot of these people would tell me that! So I refused to go until 1948 or '49. Then we got married, because the law changed. And we got married in '49 or '50. I don't remember which one.

Cleveland came from Hope, Arkansas. But he contracted, all that time when I met him, he be haulin' Indians from the reservation and bringin' them out in the fields, pickin' cotton mostly. And then, after we got together and had our kids, then I started helpin' him. I'd drive a great big truck, and go to the Papago Reservation. We'd bring out Papagos to pick for us and chop for us. Then maybe we'd take a trip to Arkansas. I went out to Arkansas about four times.

Then we'd bring families, maybe five or six families, back. But they just came with their clothin'. That's all. So we bought them stoves, tables, beds, dishes—what they needed for each house. And that's the way we took care of them. And then we bought them groceries. But they'd have to pay their grocery bill back to us when they made money. Most of the workers were Arkansas people, just a few from Oklahoma. That's how we started, and that's the way it went on until we bought this place.

The contracting ended in 1960. He was still doin' it when he got sick. I think he started contracting, in '44 or '45. I don't know exactly how contracting works. . . . You just go and talk to the farmer, but can't nobody else come in when you say you gonna do all the work. You gonna take care of all the fields, do your choppin', do your pickin' and everything. You cover it all. And then you put it in writing, that you are gonna be responsible and do it. But if they trust you, you're fine.

So we'd do that for, like Mr. Ladd, who was a man here. We contracted for him. Mr. Anderson was another one. We contracted all that field on the back side where all them trees at. Way on down by La Palma, on across the railroad, and then on going toward Casa Grande we contracted another field for Mr. Devon and Mr. Snitzer, and then Mr. Henry on that side of Casa Grande. And over here, Mr. Wadell, and Mr. Laughlin and others. It's a lot of people that we worked for, some I can't even remember their names.

The farmer would have to pay us. And then we paid all the people. We used to chop a field, maybe an acre. We'd get fifty cents off of each acre that we chopped. It wasn't much, but it was a lot in that time, you know. We made a livin' out of it, but it went higher later on in years. Then you had to chop so many rows to an acre and then step the field out. He does all that. You have to see how wide, how many acres in there. Because sometimes you don't really know and you'd have to walk the field. He was good at that. He'll do all that and then space them off. Maybe this person would say, "Well, I'll take two acres." So we'll step 'em out, you know, give 'em maybe eight rows to an acre. Give 'em sixteen rows. And maybe it be the longest, from here to the highway over there and that may be two acres. Maybe they make eight dollars at that time. Four dollars an acre. That was hard work. And some are fast choppers and some are slow. So they'll make three acres at the most.

Sometimes they paid them by the row and sometimes mostly by the acre. Then later on they went to payin' them by the day. That was better. To me it was. Because you don't have to rush, you don't have to be tryin' to be in a hurry to get your acres out, or make a mess. So they started payin' them about seven or eight dollars a day. Now they get thirty dollars a day for choppin'!

When you picked cotton, was it two dollars a hundred pounds, or three? But anyhow, we used to have Mexicans come out of Texas that had nothin' but mostly girls. One family had five girls. And the father was standing in the middle and put his daughters on each side and the mother she just kinda fiddled around in between. And they picked them rows, they'd be singing up a storm, and they made good money. Same way with the blacks. The blacks get out there and sing spiritual songs and it just be their sacks in between. They'd be singing, just pickin' cotton. We'd have to crawl up on the trailers. There would be a big board across the top of the trailer and you had your sack on your shoulder then dump it on the truck and the wagon. Come back down and go again. We didn't have no certain hours. We'd go real early, when it was the cool of day, and just work until dark, then come home.

Myrtle Jordan, with migrant workers and some of her own children.

As far as the average payment for the workers or for the contractor, I never did know too much about how that went. I don't know why I never did question my husband about that part, but it's about the same as chopping. The contractor gets so much out of each hundred pounds. And you get so much a bale, when you pick a bale, it's so much. Good pickers make over a hundred dollars a week, and that was good money to them. It depends upon how you pick, because you have to work hard at it. Some is fast pickers and some are slow.

The black people were fast. Oooowe, they'd pick fast! Some just seems to me they enjoy it, but a lot of 'em make good money, but in them times they did a lot of gamblin' and I didn't know nothin' about gambling. I've never seen that before. Another thing that kind of got me after I got married is when we went to Eleven Mile Corner. He brought in a truckload of people. Some were jet black but had the prettiest black hair, you know. They were jet-black people, and I've never been around blacks, and I was kind of uncomfortable. . . . I got acquainted with one of the ladies; well she was a girl at that time. She was jet black! Like I said, carbon. She had some long pretty hair. And beautiful—she was just as pretty as she could be. Her name was Orangelle. I don't know what ever became of her. But her biggest problem was gambling. Ooooowe! Shootin' them dice. You can hear her from Friday night, Saturday night, Sunday. And Monday she back to pickin' cotton. Some of them they saved their money; some of 'em . . . well, they gambled it off.

Now Orangelle, she used to bring her money to me. She used to call me

Myrt. She says, "Myrt, keep my money for me and don't let me have it if I come back." And when she comes back, I wouldn't give it to her. Oooowe, she'd be so mad at me. I said, "You told me, you know what you told me?" "No!" Oh, she'd be mad at me. 'Cause she said, "this is for my savings. I don't wanna touch it. So I'll have money to go home." When she did leave, she went to California. She didn't go back home. She didn't have no children, she was single. She was just a young woman.

One day we were eating supper—we had some big 'ol cottonwood trees, it was real cool there, right by the canal—and these women were sittin' out. They had a fire, you know, they was ironin'. They had them straightenin' combs on the coals there. Then I saw this woman's hair smokin'. And I said, "Honey, that woman's hair is on fire!" And he [Cleveland] just started laughin'. He said, "Baby, that woman's hair is not on fire. They are what you call straightenin' their hair." I said, "Ooooowe, it was just smokin'." I said, "What is it?" He said, "Come on, let me show you." So I walked out there with him. And he was laughin'. He told that lady. He said, "My wife said your hair was on fire!" "She never seen a nigger's hair on fire," he said. I said, "Well, I'm sorry, but I didn't know." So she said this is the way we straighten our kinks out, you know. You have to do this. I'd try to comb the girls' hair, and I couldn't do it. My children had "good" hair. They had little curly hair, but not kinky hair. But the boys, these younger boys, Ronnie never had it, Berdette did. After they grew up, their hair changed and they wore big Afros.

After the contracting went down, the people mostly went to California. And all the Papagos went back to their reservation, but some went to Maricopa and on around and just workin' on farms. A few of them did. I think that's when crime got real bad, 'cause there's nothin' to do, really nothin'. We used to have a yard full of kids every Saturday. We'd have a yard full and load 'em up on the trucks and take 'em to the field and they'd make a little money just to go to the movies or whatever, and we'd bring 'em back, by one o'clock.

Saturday, see when we take the kids, we have a grown-up to watch them and weigh their cotton or whatever. We had two trucks, he'd [Cleveland] take one truck one way and I'd go the other way, and load up some of these Indians and take 'em to town. And he'd take a load to town. And that's another big problem. You'd say, we gonna leave at such and such a time. Oh, it's so hard to get 'em out of town. Some get to drinking, then a lot of times we have to leave. You know, just leave some of 'em and come on back.

There was one bar-type place right here in Randolph, but that didn't last very long. Some Mexicans opened it up. But it didn't last long. My son Berdette used to play in the band, and he used to be over there poundin' his drums.

In La Palma, they used to have a bar, and they had a lot of dancing and a lot of that kind of stuff. I've NEVER been in a bar. I've NEVER went dancin'. I've NEVER been really with the Indians, and so I didn't know. They asked, would you take us to La Palma? I said, "Well, what's wrong with you?" So anyhow, I take 'em and I tell 'em, "Well, I'll just leave you all here and I'll be back later on to pick you up." When I do go back, no Indians, where are they? [Laughter] They all went different ways.

We had camps over here for our workers. There used to be a lot of little houses. We'd put some Indians there, and on the farmer's ranch and Ladd's place, we had some. So, everybody had a place to live. We had a lot of black boys there and one bunch was always slow gettin' up. They'd be out there cookin', ya know, and they'd have hoecakes or somethin', they'd be flippin' the pan and says, "Oh, we've got to go. Load up!" Then they'd say, "Wait, I've got to flip my bread!" He be flippin' his bread. Oh, I used to enjoy that. I didn't know nothin' about greens. I ate spinach, but a lot of things I didn't know I learned from the black people. My husband taught me a lot that I didn't know. I never ate a pig tail, or a pig foot, in my life! I never had fixed Indian food neither 'cause I didn't grow up on the reservation. I remember seein' my mother and my great-grandmother and my grandmother make things, but I never learned because I didn't grow up on that.

When I met my husband, his father was living here in Randolph. They lived on the highway down here. His sister was livin' here too. And I didn't know nothin' about Randolph. I didn't even know there was a Randolph. And he brought me down here and introduced me to his people. I was scared to death. I'd never been around black people. And it just, well, they talk different, it was just somethin' I couldn't hardly understand, and I said to myself, "Now what am I doing with this man?" Ooooowe, I couldn't take this!

I didn't know anything about black people. I seen them, and the first black I got close to was a little boy that went to school at the Indian school. And he had real nifty hair. Real nifty hair. And he was in my class, and we used to walk and touch his head. His name was Stacey. And he'd say, "I wish you girls would quit touchin' my head." And where did he come from? I don't know. Was he that Indian woman's son? Where did he come from? But he went to school there with us, and later on I heard he married an Indian girl from Blackwater. But that was the first little black boy. He wasn't black, he was a light-skinned little boy. But the hair was real wiry. Just kinky.

When I first went down there where we were supposed to live, I had real long hair. Real long hair. And my hair was brown, it wasn't black. Just brown, kinda reddish brown. And them black women would say, "Ooooowe, she got beautiful hair, such a sight." They would brush my hair, and I got tired of brushin' my hair. They just teased my hair and just feelin' in my hair, and a lot of them black young boys like, "Ooooo, you're so pretty."

As to my second marriage to Cleveland, my dad and mom didn't mind, but my grandparents, they talked about me. They really disliked that. Didn't want nothin' to do with it. Just 'cause he was black, I guess. They didn't want no black person in their family. No way! I went on and got him anyhow. Then when I had Ron, well, I took him to our church. I had him baptized in my church, and he was the quietest baby there. He didn't look black at all. Berdette didn't look black either. None of my kids looked black when they were small. Except when I got way down to Moochie. Moochie was the darkest kid I had.

Cleveland and I had twelve children of our own. Let me see. I can' keep up with all of these kids. There's Ronnie, Berdette, and Burton, Zelda, Daryl, Sheffield, Moochie, Henrietta and Virgil, Jeffrey and Adrian. A daughter, Marilyn Joanne, died. It's twelve, right. It's twelve children of our own. And then I took in eight of his kids—Grover, Georgie, Pearl, and Bessie early on. Then Willy, Bennie, Grace, and Ada joined the others after their mom could not care for them any longer. So it was quite a bunch. But by that time, Ron and them was growin', and some went to school in California. He was in Compton College at that time. And so, we didn't have all of them at one time. After they grown up, they left.

We bought this place here durin' that time. You could buy quite a bit of property pretty cheap. Now you can't do it. I think we bought this place for five thousand dollars. And it's from there clean on to the back side. But this little house was here and was nothing but that little bitty part. That was all. This wasn't here [pointing to an addition to the house]. Right up to that window. Just that little house. When we got it, it had two little bitty rooms, kitchen area, no running water, and the front room, a little hallway, a little closet. It was real tiny. And this little room and that back room. And that was all. And we had all them kids. At that time I think we probably had about five at that time. And then, and my stepgirls, he [Cleveland] had some children, you know. So I had them all.

So we and some of the children slept on the floor, and then we built cabins around there. About one, two, three, about four, five, six, about ten, twelve cabins. And the boys stayed in the back in a big old bus. It's the body of a bus. But we made beds and little tables and stuff and made a little house out of it. So part of them stayed over there. So, that's what we did. We just worked and worked and kids went to school at Kenilworth. Some went to Eleven Mile Corner. It's the oldest two I think that went to Eleven Mile Corner a very short time. Then we moved over here. So then they all went to Kenilworth School.

That's my grandson there, that just came out of the house. That's one of my daughter's. And I've had him here since he was five years old. We call him Bogie. He's goin' to work. He works up at the Training Center. He's twenty-four, I think. He was born in '66. I've got about fourteen great-grandchildren. I don't know. And grandchildren. I got forty . . . last count I had was forty-two or forty-three or somewhere in that area. I got three boys. The oldest boy is in California. Two little white kids over there. Ron got two, Berdette got three, Burton got three, Daryl got three, Sheffield got two; my oldest son, he had eight children. Duane got four. And Adrian, he got four girls, and Virgil got six, and Zelda, she had two, the mother of this boy and his sister.

My husband, Mr. Jordan, was a hard worker, but he'd discipline his kids. He used to tell them that, "Well, you gonna learn the value of your money. One of these days you might thank me." And they work on the weekends, he'd give 'em their little pay, you know. We didn't have much. He'd just tell 'em, "Eat what you can, but don't take nothin' away." He worked hard and he provided for us. We didn't have to walk. We always had vehicles, two or three of them around. He was rough on his kids at times. We never had a dirty yard. We never had a dirty house because he'd make 'em clean up, and when you get out of the bed in the morning, you had to make up your bed and you didn't get in it unless you're sick or sleepy. No, you don't bother the bed. You play outside. You get dirty, you get in a tub of water. [Laughter] But he made 'em work. He sure put 'em to work. Make 'em pick cotton. Make 'em chop. Made 'em work hard. They hated that. They'd say Dad was mean. And I think that's the reason why all my children turned out to be like they are today because they were really taught the hard way, you know. They had to learn the hard way, how to work. You don't get nothin' by just sittin' around. And so I think my children did get somewhere on that part. Sometimes he would take the money they made on Saturdays from them and save it for school, clothing, or something.

We used to have cows out here and chickens and stuff, and that's another thing the kids would have to do. They'd have to get their buckets and go out there and feed the rabbits and chickens and milk the cow. We had a cow we had to milk, but my oldest stepson, Woostie, could milk her, but nobody else could. She was mean. And I couldn't milk her, and my husband couldn't milk her. But shoot, I didn't have time, so we had problems with that old cow. We had hogs, they had to feed them pigs and water them; and we didn't have running water at that time. We had to haul the water. And that was pretty rough on them kids to do all that. So they'd be whinin' and goin' on tryin' to do things. They'd get it done, but it was just hard work.

When we bought this place here, we had a pump out there. There was running water. Later on, I guess, from so many people pumping, it just sanded out. We tried to fix it—we couldn't get any more water, and sand got built up in there somehow and choked it up. It cost us so much money we just quit. We just started haulin' water, and then later on we finally got water. Ooooowe, that was a great thing! Water all over, we had to pipe our own, had to pipe it over here.

I really don't know how many acres we had, from this side of the railroad, to that gully there, right up to Jeff's fence. But I sold all that part over there not to the Mexican's house, that part wasn't mine. We bought that down on that end. Then another Mexican bought a little place before my husband died, way on the far end, too. So we just got a little strip. Just a little piece. I don't know how much is left.

After we got our water, I had a beautiful garden out there. I had greens, tomatoes, peppers, okra, just all kinds of cucumbers. I even had cantaloupes out there.

So that's how we wind up here. We just contracted field after field. When I had all the kids, I just helped them haul the workers to the field and leave them, help them out weighing cotton. Haul water too. We used to have to buy ice by the blocks and put it in the barrels for drinking, and then I ran the lunch wagon. Oh, it was hard work. When my husband had a stroke, that's when I kind of made up my mind; I think, "I better find something else to do." So I asked him if it was OK if I go over there to the Training Center and apply for a job. And he said, "Sure."

So I went and applied in '64. He died in '65. I went over there, to the Arizona Training Center, and applied, and two or three days later they called me up to come to work. So at that time we had to go through a physical and stuff and wear nurse's uniforms, so I had to rush and get organized to go to work. I was there for a little over twelve years. I worked at the infirmary, where the little sick kids be at, takin' care of all the sick ones. And then I went to the other part which had severely retarded kids that lived in the beds and can't help themselves mostly. But sometimes I'd give medication and treat them, you know, wherever they need to be treated, such as sores, or just take care of them. And other times I'd just be on the floor supervising the people.

The families that we brought in to contract have all moved on, now. There's no one here. And I don't ever go to the main part of Randolph, but I know the people there. We've been around each other all our lives. The Whites and the Moores . . . Miss Fannie; we've been all here; we used to visit a lot. But that's all stopped. I don't know why nobody does this anymore. The Fields used to come over, and we'd sit out here and talk. They done got old now. I don't ever see them. They don't come see me. Sometimes, when all my kids leave, there's no one here, like this boy gone to work. Just me. And lot of times, it gets pretty lonely. So my son here says, "Well, Mom, why don't you find you a little ol' man? Hold your hand, if nothin' else?" I said, "Shoot, if he had money, he sure can hold my hand!" [Laughter] Yeah, he sure can hold my hand if he had a little money in his pocket.

But it's true, my children are real close to me. This one lives here, but they all come over if they have to. When he's hungry, he's over here. "Got anything cooked?" I'd say, "Yeah." And I'd tease . . . sometime I start walkin' across and he hollers, he say, "Mother, what you doin'?" I say, "I'm comin' over." "Ah, you ain't comin' over, I'm leavin'. Just turn around and go home." I'll turn around and come on home! And then when I see him, I say, "Go home, I don't want you over here." You know, just little things, but Ronnie and Burton are in Phoenix, and this boy and Sheffield, they're all the time over here doin' somethin'. Sheffield mowing my grass, you know, just little things to show they care. Burton put a door on my bathroom which was about to fall off. And then Ron worked out here and we had a family gathering here on the Fourth of July. They had a volleyball game and we had tables out, just a lot of kids and grandchildren. Oh, I got black ones and red ones and brown ones, you name it. And we see each other all the time.

I don't have but a few relatives on the reservation now, and I go back and forth to go to church over there. I have two brothers in Blackwater, and my sister's down in Sacaton. That's all I got. I go and visit quite a bit, but it's a distance. My sister and I, we're pretty close. I don't know my older brother's children. Somebody might walk up to me and say, "Hi auntie." And I look, and I'd say, "Who are you?" Yeah, I'd have to ask him, well I'd say, "I'm your aunt, who are you?" Then he'd tell me, "I'm Clayton's son" or daughter or whatever. I'd say, "Oh, my God." Yeah, I don't even know them.

I go to church in Blackwater, that's on the Gila Reservation, and the Pima too. Well, it's all one reservation, but it's different names, you know. Like Blackwater and down below, Sacaton Flats and then Upper Santan, Lower Santan, Sacaton, Snaketown and Maricopa and whatever. Then there's Ak-chin, and just Little Gila Crossin' and Lehigh Levine all goin' that way. The house where we grew up burned down when we were young. Then I left home after that. It's not there; it's an old shack there and it's trash all around. But the Indians, they're not very clean people, you know. They don't care about pickin' up stuff. They got nice homes, but they don't take care of them. I wished I had one of them houses. I could go back but, no, I couldn't, I don't want to. I'd never want to go back again. I'm just black, that's all. If I was to marry again, I think I'd marry another black man. Yeah, I would!

My son Moochie considers himself black, but Ron, he's more Indian. He works with the Pima Indian Tribal Council. Ron is very, very into the Indian thing. He goes to Sacaton, where a lot of them Indians have this council meeting. They all sit there, and they talk a lot when they're on the outside, but you get 'em together in a public place, they have nothin' to say. They just sit there. The last time Ron went to a meeting on the reservation, my sister-in-law said, "Nobody speaks out." But you know Ron, him and one of my nephews, they are the only two that stood up and just told 'em what they thought about whatever they was discussin'. Just sittin' back there, even the big shots just sit there, wouldn't say nothin'. And Ron don't even live on the reservation. He stickin' his neck out for them Indians.

I never want to go back to the reservation again, as I said, I'm just black, that's all!

interesting chapter re race relations and local generalizations

The Jordan children and grandchildren at Cleveland Jordan's funeral in 1965.

Jordan grandchildren and great-grandchildren, July 1992.

Jeff Jordan

Jeff Jordan, the eleventh child and eighth son of Myrtle Williams Jordan, is now in his thirties. He lives next door to his mother, and on many occasions, as I visited her, he would show interest in what I was "writin'." He urged me to find out what was going on at the present time in Randolph and the surrounding communities. He struck me as a rather "angry," concerned, and socially conscious person. Our conversations about the state of blacks and the underclass were always filled with passion and sincerity—his voice is a mixture of Tom Joad from The Grapes of Wrath *and Jesse Jackson and Nelson Mandela, all those who showed impatience with the reluctance of the power structure to do something about the inequities in the systems that subjugate the powerless. Jeff, like them, "saw wrong and tried to right it." Like his brother Ron, Jeff sees the need for a public conscience and for minorities to design their own destinies.*

The Sunbelt Refinery.

The Proler recycling manufacturing plant is about one-quarter mile from the Jordan home on Kleck Road, and its fumes, for years, had bothered them. When the Sunbelt Refinery opened up to the northeast of Randolph, Jeff and the community were boxed in by the two plants and their hazardous fumes. Jeff saw this aspect of the "new" Randolph history as extremely important to document. This unexpected development would again change drastically the face of an already declining community. Other similar environmentally unhealthy incidents were happening throughout the state; the two other early migrant communities of Mobile and Vicksburg, to the west of the freeway, were having similar experiences. They are all unincorporated towns with no political representation. Besides these three communities, the Pima Indian people on the Gila River Reservation, especially the community of Blackwater, are also in the plant's odorous path. Clearly, it all smacks of environmental racism.

Jeff Jordan, as the grandson and son of Arkansas migrants, has a notion of what the community's life and time were, are, and could be. He feels that his generation must become vocal and activist to make things better for the future. Although we talked about many things from his childhood to the present, it was his rage about the refinery and what the powerful are doing to the powerless, the underclass, and the generation he represents that was loudest. His second major concern is the fate of his generation and those younger than he regarding jobs, education, and opportunity. Like many other Randolph residents, Jeff has the gift of leadership, common sense, and a missionary zeal. Perhaps with voices like his, this town may yet rise again despite the overall lack of optimism.

His narrative reveals a tired and worn-out Randolph, struggling to remain on the map. It is what you see, now, not far from Selma Highway, the road that runs from Eleven Mile Corner to Highway 87 at La Palma. It is not a highway to freedom.

* * *

I've heard different people talking recently, about the refineries, and there seems to be a panic going on about industry in the area and what it's doing to people's health and property values. You seemed to be involved in the movement to investigate their presence in Randolph. Can you talk some more about that?

Well, they showed up, I believe it was in 1988 or '89. It was when they really started pushin' it. And when they did finally get it through, seemed like it didn't take 'em very long to get it built, complete and functioning. It was a rapid, rapid completion of a job. This is definitely having an impact on the community, without a doubt. Me and my wife was just talking about if things don't get better within another year or two we may have to move. Simply because of the air quality there. It's pretty bad. So I was kinda involved in trying to get somethin' done through the court system. Myself and Melvin Moore and a few of the other residents in Randolph are tryin' to get some action taken.

Well, the Refinery is not even a quarter of a mile from some residents in Randolph. Maybe less. And my thought is that the whole thing was done wickedly, politically, for greed, for money. It's inconsiderate for other people, for human beings, to put somethin' like that here. I was never told that they were gonna put it here. No one asked my opinion, what I thought about it or how'd I think it was gonna affect my children. I think the county did it, kinda underhandedly toward the people of Randolph.

If you consider that in Randolph most of the people is probably in the poverty income level anyways, probably 95 percent of 'em are, and so—this is just my thought, it was just a theory. They thought, "Well, what can these individuals do? They don't have any money." They're in the poverty income level. The state more or less almost owns them because probably a lot of them are on public assistance in some way or another. So, what could they do? And it's not that many of them, you know. There's not a whole lot of people in Randolph.

If there was like two thousand people, then it may be a little different. But it's probably only a couple of hundred. And that's including the children. If I actually sat down and calculated, it's probably less than two hundred. Maybe one hundred. It's just a very small community, but I think if we got good counsel, a good representative, or attorney, I think they would have a good case, a very good case. As a matter of fact, I cut an article out of a magazine yesterday of a lady back in Louisiana that had a case similar to ours. They had several refineries in their community. Her name was Janice. They won a class-action lawsuit, and they didn't keep their homes, . . . or they may have kept their homes and the property, but the government or the state relocated them. And I'm kinda hopin' that perhaps somethin' is done like that for the community here. They didn't relocate the refinery, they relocated the people. That's hard because you're moving, and uprooting families and your life . . . you're losing everything; you're losin' your cultural ties, you're losin' everything that you know. You take a lady like Mrs. Reece, the elderly lady. She's been here probably fifty years, maybe longer. They gonna run her off! She's not a very healthy woman. You take some of the other elderly individuals like Mandy Moore's mother. She told me the other day she thinks she's being affected by it all, by the Refinery.

And that's the reason I'm so mad, because it hurts me to see the people who have been here for forty, fifty, myself thirty years, my mom probably fifty, and we being run off our places. That's not right. I think it's gonna be a hard struggle for us, though. It hasn't been easy for the people here in Randolph, and it's gonna be very difficult, I think. What I mean by difficult is getting representation. Because nobody wants to work free. That's the reason why myself and Melvin is talkin' with these different attorneys. We just waitin' on a response from them to see if they would handle the case.

Me and my wife, we in a position where we can leave. We don't have a lot of money, but we have money where we can relocate ourself, but a lot of them can't. My mom can't. People on fixed incomes can't. The elders, you know, what are they gonna do? It would make sense to try to get the business out, not the people. I don't know what's gonna happen. But we haven't given up. We thinkin' about organizin' and different things, like a protest march, lettin' the refinery people and the newspapers know how our quality of air is here, how it's affecting our elderly and our children. People are complainin' about headaches. They're vomiting, the whole nine yards. And it's affectin' the individuals, and some of the people in Coolidge and Casa Grande and Phoenix don't know about this.

Well, that could put Randolph on the map because the Tucson papers and the Phoenix papers and so forth, even the Casa Grande paper, all get to a lot of folks; then they would know about us. They would know what's goin' on. The Louisiana case was similar to ours, almost identical. They had another industry come in, and they referred to that street as "cancer alley." They won in the sense that they got away from it, but they lost in the sense that they lost their heritage and their homes. And even now, our property is goin' down. No one would buy your property here. We got people just on the other side of Randolph that probably own $100,000 homes. People just west of the refineries, them individuals got homes probably in the range of $50,000 to $70,000. They got homes that are expensive. They probably couldn't sell their homes if they wanted to. We got people within a mile, probably a mile and a half radius of this refinery that are smellin' the smell. Not as heavily as Randolph, but they're getting it, particularly like on a windy day.

This county is broken down into many districts. Each district has a supervisor. Right now [1992] they have an election year. The supervisor for this area, he's no longer runnin' anymore. And so, there is two new guys that are runnin' for this district that I know of. You go along Highway 87 you'll see a supervisor of District One. The guy that's presently in there, I haven't seen his name. Then I heard that he wasn't runnin' this year. The county supervisor had to approve of somethin' of that magnitude. The only ones that are really benefiting from that refinery is Sunbelt themselves, the county, and the state. The only one that's gettin' the tail end is the people that live nearby, and that's Randolph. The present supervisor for this district is no longer running. I don't know how long the terms are. He was probably the supervisor for this district for a three-year or two-year term. There's a couple other guys that are running for that position. They're the ones who represent District Number One. And we're in that district.

I don't know if you've been watching the papers, but I've been picking this up for the last month or so. All over Arizona, there are a lot of supervisors resigning. They're not running. Roy Laos, who's from Tucson and became a supervisor in Marana, which is about twenty miles up the road, he's not running. The city of Tucson supervisor resigned. He appointed somebody to take his place 'til they find another person. We're not getting the reasons for these guys resigning. I think there's some big complicity, and on election year it's gonna come out. It could hurt the bigger election people, like senators and presidential elections and stuff. It's something going on in Arizona.

It's interesting because the refinery's in the county, Pinal County, the county we're livin' in. The county doesn't have anything to do with the refinery. Isn't that odd! Because they approved of it to be in Pinal County, but they don't control the quality of air at the refinery. The state does. I don't know why the county doesn't control them. The reason I think so is because they don't have the professionalism to do that. They're not qualified to handle somethin' of that magnitude. Pinal County is probably benefiting from the revenues for bein' there, for permits and things like that. But other than that, they don't have anything to do with it. But the bad quality of air is here in Pinal County. It's in Randolph. It's in Twilight Trails. It's in Martin Road in Coolidge. It's all the way out to Wheeler Track. It's people from La Palma and the Arizona Training Center. And different places like this.

I don't know of anyone from Randolph that works at the refinery. From all the guys that I talked to, and even some of the women as well, no one showed an interest to work there because of the fumes. See, that's what we're fightin' right now. The fumes and the gas odors that's comin' from there. Who would want to work in those environments? People would be a fool to work there. You know, if you had a clerical position, or if you had a supervisory position where you're inside, it wouldn't be that bad, but who would want to work out in the field over there? The better jobs are probably the jobs sealed off from us. And not many minorities are probably gonna get those. And if you look over there or just drive over there or walk along the track and look at the refinery building that the people are workin' in, that building is solid. The people that are workin' there, they don't get the fumes like we do because they workin' in a very sealed tight area.

A lot of the big-time senators, they're tied in to that. They want it there. And so, they close their hearts to you. You know, you go over there and talk, "I'm gettin' some pollution from this refinery. Could you help us do somethin' about it?" The refinery refines gas for different Air Force bases. All them guys been in the military and they're military men or have the background and things of this nature, so they're keeping their mouths shut.

We signed a petition and we submitted one to Air Quality Control in Florence. We submitted one to the county attorney's office in Florence. The attorney general's office in Phoenix got one, and I'm sure these individuals were aware of the people's complaining about the air quality in this area. It's been on the news already about them bein' in violation of so many different things. That spot was strategically selected because, as you said, powerless people in the county don't have control. Now it's very difficult to shut it down because there's so much political power involved. And so, the people themselves are probably going to fight for relocation. At least, that's what I'm goin' for. Every day, five, six, seven truckloads, delivering oil, taking out gas. We can't live here. I got a five year old, a nine year old, a fourteen year old, and what is my kids' health gonna be five more years from now? We drive down the road, and they've got their hand over their mouths and noses. "What is that smell, Dad?" I tell them, "There it is, right there!" It's not like it's two days out of the week. It's every day out of the week. It depends on which way the wind is blowin'. But it gets in your house, it gets in your car, it gets on your clothes, it's everywhere. It gets on you too, Geta, since you're here a lot too.

Randolph may eventually want to industrialize. I know they talked about closin' the Training Center. If they shut the Training Center down, it's just givin' prime reason to do so, because the people here will have a complaint. Perhaps in court, we can try to do something for relocation or compensation.

You know, there was a great plan for Randolph about 1925 to make it a resort area and build it up, and they were sending flyers all over the country. This is the place, they said, it would be bigger than Phoenix. And maybe this is it now. Randolph has arrived, almos' after seventy years! If you look at this area too, it's just kinda ideal, because nobody wants big-time industry in Phoenix anymore because of taxes and all kinda requirements that these companies have to meet. They don't want 'em in Tucson. Environmentalists are very strong in these areas. Very strong. They got all kinda propositions on the ballot the end of this year for people to vote on. You got mostly people that are at the poverty level, particularly here. Even Coolidge is falling apart. There is more and more closings of businesses. The streets are dead, I mean just boarded up. So that's not coming alive. So this would be a place not that far from Phoenix. Casa Grande is doing its thing over there. That may be the grand plan. But I think in the future, without a doubt, it's probably gonna change. I give Randolph probably five, ten years, and it won't be a Randolph anymore.

We don't have money. We're kinda in a situation. Well, we just bein' oppressed, constantly oppressed by the way this system is. I know for myself, I wouldn't want to stay here another two years. I built my addition on my house simply because I needed the room, and it's not gonna take a whole lot of money because I'm doin' most of it myself. If they should shut the refinery down, I could always come back or I can rent it out or whatever. What hurts me the most is, here I spent all my life, my home, my property, my children, and it's all been washed away.

attachment to place

Well, that's not the only thing affecting us here in Randolph. The young people are a concern. I think the biggest problem affecting them is the fact that they can't get any jobs. Right now, work is the biggest major problem for a lot of them. There ain't a whole lot here. You look on the farms, most of the farmworkers are pretty well taken because we live here in Arizona. A lot of illegal aliens come here, and they come to work. A lot of these farmers hire a lot of illegal aliens but not the local people. They don't have the skills for farming. Coolidge only has a population of probably five thousand or six thousand. Casa Grande ain't too much bigger than that. Maybe fifteen thousand and twenty thousand. And there's not a lot of industry there either. A lot of them don't have any special skills. Some of them barely read well, so they have no skills to go to the big places. Beyond Casa Grande, they can't even go to Phoenix and compete well, or California or somewhere else. So they kinda in a little spiral. Some of the kids have opportunities to be athletes. That was one way out, but they gotta have that schoolin' too. But then, everybody can't be an

O. J. Simpson or Michael Jordan. You gotta have the skills, the right body, and coaching and all that. Not every kid would be a super athlete. About 2 percent make it to the top. So that's not an answer.

There's not very many of my classmates here anymore. There's a few, but not that many in Randolph anymore. Barbara is still here. All the rest that I know are gone. Deborah's no longer here, she's in Coolidge. Most of the others are working at the Training Center. Tommy works as a bank teller in Coolidge. But some of the other classmates from Randolph, like Annette, she's gone. She's in California now; her husband's in the service. Sue, I'm not sure where Sue Talbot is, or Ann. They're no longer around here. Tammy's husband passed away. The young ones, some friends of mine that could leave, did. I don't blame them. Some come back to visit, now and then. What you're seeing now in Randolph today is a few of my classmates that are still there, not very many. You see a lot of the elderly that's been there for forty or fifty years. You see, perhaps, a few Hispanics—individuals that come in that are farmworkers. I've seen some Hispanics arrive in the last two years. I'm seeing trailers right between older homes. There's plenty changes.

I'm part Pima Indian on my mother's side and black on my father's, and so I was always kinda concerned. Not that I lean more toward blacks than Indians, because I don't. It's just that I recognize that black people were oppressed more, particularly in these areas, than the Indians were. I can remember a sign in Coolidge when I was a kid. It was a long time ago. There was a sign that said, "No Indians, No Mexicans," and then it had another sign that says, "No Dogs." And they were talkin' about black people. Yeah, they weren't talkin' about no dogs. And they were referring to black people as dogs. It still happens today. It's just not fair; the signs are gone, but the hatred is still there. It didn't get any better, you know. Randolph ain't got long.

III. "Next to You, We Love Cotton Best!"

The Fabric of Black Life

> "I never did pick five hundred pounds. . . . I never did get to five hundred! Now, my wife picked five hundred pounds every day. . . . When I got 492 pounds, all I had to do was go back out there and pick that other eight pounds. But I said, 'I'll do it tomorrow.' Tomorrow never got you. So I never did." –Nick White

No novel in the twentieth century has stimulated debate about the plight of the white migrant poor more than John Steinbeck's 1939 *The Grapes of Wrath.* We read about all those poor Okies, symbolized by the Joad family, in wretched vehicles loaded down with most of what they owned: chickens, dogs, beds, pots and pans, children, and each other, takin' to the road in the Depression years of the 1930s. They broke our hearts in the novel and the award-winning film. All of them were on their way West to seek a better life—never looking back.[1]

James Agee's *Let Us Now Praise Famous Men,* also published in 1939, chronicled the white Appalachian poor with poetic words and voices, and photographer Dorothea Lange's work captured, vividly, the places and faces of the migrant poor in the Casa Grande Valley.[2] We must consider, however, how the white Okies' journey would be different from that of a black Okie, Arkie, Kansan, Texan, Missourian, Mississippian, Alabamian, at a similar period in our history. Their journey began long before, when they sought sanctuary from slavery's whip and the lynchings that followed Emancipation and Reconstruction.

The journey of America's Negroes from slavery to freedom began when their hopes for "forty acres and a mule" vanished, as land remained concentrated in the hands of a few and the Black Codes clearly favored white landowners and employers. It became evident that to support themselves and their families, many would have to return to working the land of their former owners or of landlords who made similar exploitative demands on their labor. Many suffered severe disillusionment and sorrow.

We must consider also that one of the characteristics of "free laborers," as contrasted to slave laborers, is that "free laborers" work only for a limited and specified period of time, after which they are freed from the domain of their employers and free to govern their own lives. Correspondingly, the freedom of such laborers is predicated on their mobility within the labor market. This means the freedom to choose their employers and the freedom to leave a given sphere of production when the prospect of better opportunities arises elsewhere. Formally, all "free" labor entails these freedoms. In practice, however, these freedoms can be limited both by the market's "hidden hand," which determines the availability of employment and/or the price paid for labor, and by the fact that people without income are compelled to work for others to survive. Indeed, under the southern systems of slavery and sharecropping, these freedoms were severely limited. In spite of an edict that granted black freed men the right to move on, the freedom to seek and maintain work no matter how menial or noble was always fraught with difficulty. White landowners, having lost their free labor, lay in wait to exploit the hungry "New Negro." In their eyes this New Negro was no different from the old. A piece of paper meant little to them. These newly freed people, recognizing that Jim Crow and Massa were not "a-moldering in the grave," took the roads going North and West, hopin', prayin', and walkin'—to freedom.

The Great Migration, as we have come to know it, began between 1915 and 1918, when 300,000 to 500,000 blacks left the plantations and moved from the South to northern territories seeking jobs and equality. Although some of them found the jobs they sought, they never found the equality. There were many agricultural jobs vacated by soldiers during World War I. There were also fertile lands. As these blacks migrated northward, at their first stop, in southeast Missouri, they found swamps being drained in anticipation of bumper crops of cotton. Some stopped there, believing themselves far enough away from Mississippi and from southern hate and dangers. The January nights were so cold that even the oldest present extended a fragile arm as if to touch the flickering flames of a burning campfire. Much of these areas resembled a barren wasteland. The abused soil was barely able to support a seasonal crop. Grass had difficulty sprouting. Potato sacks, blankets, anything that might be utilized to protect one from the cruel cold was put into service. Often a fence provided four walls and the security of a living room. A warm fire in an old wash captured the warmth of a winter night. Many people huddled close together to keep warm. Race was sometimes easy to forget in the face of mutual struggle. People remained fully dressed in the cool of a sleepy night, only to be covered and harbored under the gentle wings of a morning frost.

Even a feline friend weathered the storm of the chilling weather. All of their belongings were wrapped in sheets and blankets. All that one owned, all that one could muster from an entire lifetime, was present in these heartbreaking scenes of blackness.

A young man recalls his grandparents talkin' about those days and how their words instructed him about that part of his history and the personal exodus of his family at a time when the newly released, landless, foodless, and homeless black people took to the road. He says,

> My mother used to tell me that we were blessed to have a roof over our heads. I never understood why she said things like that. I thought that everybody had a roof. What was important was what you had under your roof. Our house didn't have all of the things that my friends' houses had. I envied their newly painted walls, their television, running water, central heating, and their in-house bathrooms. We did have love in our house. That lessened the sting of many of the cruel realities at the end of the cracking whip of poverty. My mother said, "Struggle is God's way of testing us. It makes us stronger, and when we struggle together, it brings us closer. My mother, your grandmother, told great stories. Most of them," my mother said, "were about her past." My mother said, "your grandmother lived ninety and some years. She was born in Carrolton, Mississippi, in 1896." So she had plenty of past."[3]

Some croppers were not so lucky as to have a blanket, a roof, or a woodstove. One said that "many of the sharecroppers didn't even have tents." It was very common to see a large group of people nestled around a blazing flame. Their faces glowed as they nipped at the sorrow by telling cheerful nighttime stories.

This arduous escape to freedom and opportunity was also a result of Nature's way of gettin' even by creating a plague that would affect black labor and white landownership. The boll weevil found its way into thousands of plantations from the South to the lower Midwest and West. From about 1900 into the 1920s this insect began to terrorize the cotton crops of Mississippi, Missouri, Arkansas, and Texas. Subsequently, the logical direction for these displaced peoples was onto the westbound highways.

In 1936, the new Highway 60 opened, and it became the black Migrant Road. Highway 60 begins in Norfolk, Virginia; it passes through Kentucky before moving into Missouri and on to Oklahoma, passing through Texas and New Mexico before winding its way up to the northeastern mountains of Arizona, then down into valleys of Apache Junction and Phoenix, up to Wickenburg, and down to Salome and Quartzsite, where it now joins the new

Interstate 10 near the California border. The road stories of families like the Joads and those of families like the Jordans and the Whites of Randolph had a sameness about them—one group on Highway 66, the other on 60. The journey West along hostile roadways and through unknown obstacles was tough. In Missouri, where they stopped to work, the living conditions were terrible, and one area was called "Niggerwood Swamps."[4]

The areas where they were housed were, at best, shantytowns. Dwellings consisted of crude shacks or tented assemblages. The shacks were constructed of wood or logs with no insulation, no plaster, no paint. They were called strip or box houses because they were square like a box, with rough timber placed together vertically. In a strip house, the spaces between the timbers were lined with stripping to keep out the weather. The box houses, on the other hand, had nothing to cover the cracks. The inside walls of these shacks, if not left bare, were covered with old newspapers, pieces of cardboard, or building paper. The floors were usually single layers of timber with cracks that exposed the earth. The only protection that many of these shacks had from the frequent floods that plagued the area was the dirt of the floors. Some houses were built on concrete or cypress blocks, usually several feet high to allow the water to flow underneath the shack. The shabby tin roofs offered little protection from snow and rain. The furniture consisted mostly of homemade items: tables and chairs, box crates, cabinets. Pallets were used for beds.

There were many health problems caused by the insects that infested the warped timber of the shacks in the summer. Other health problems resulted from cold winters. There were no indoor toilets; outdoor toilets were open and unscreened; a large number of cases of typhoid fever, dysentery, hookworm, and other intestinal diseases plagued these travelers.

If drainage ditches were not cleared out every two to three years, water became stagnant. Malaria was a huge problem. Many houses were near drainage ditches and levees, and residents had no screens on their doors or windows to keep out mosquitoes and flies. Two-thirds of all deaths in the state by malaria were in southern Missouri. This was because the drainage ditches were natural breeding places for mosquitoes. Cases of typhoid, tuberculosis, and pneumonia were also above the state average. Many of the diseases spread due to sanitation problems. It was customary to have salt pork, corn pone, and dried beans for meals. Eggs, butter, milk, fresh vegetables, and fruit were rare, as was evident from the prevalence of pellagra, colitis, and malnutrition in these areas. Food preservation was a problem in the summertime; some dug pits in the ground, used caves, or placed tubs under their houses to store food. Blacks were never housed well, but these places were as bad as or worse than the shanty townships that still dot the South African landscape.

Photographer Ken Light, in his 1995 *Delta Time: Mississippi Photographs,* graphically states the emotional impact of the cotton pickers he saw along Highway 61.

> Everywhere the eye looks, the fields lining U.S. Highway 61 are exploding with large bulbs of cotton. The air is heavy with heat and humidity, and my mind's eye visualizes hundreds of black sharecroppers being swallowed into these fields, sweating from their brows, the years of hard labor chiseled in their faces, dragging heavy cotton bags laden with "King Cotton."
>
> In this poorest of places, in this land of opportunity called America, we see the best and worst of ourselves. The Mississippi Delta is a land where ghosts of the past still collide with the reality of twentieth-century America. . . .
>
> In my four-and-a-half-year journey along Highway 61 and the smaller, less-traveled roads of the Delta, scenes that I thought had long vanished from this American land were revealed. Nothing had prepared me for this. The cypress-planked shotgun shack still sits on the edge of the plantation. Every day brings a new vision and new dismay in witnessing the lives of Delta people. . . .
>
> The dignity and traditions of the Delta's people create a powerful presence that gives hope and pride to their hard lives. Their sadness and their humanity have struck me deeply and taught me that with my freedom to see and photograph comes a responsibility to listen.[5]

Before the black migrant came West, whites and Mexicans had already arrived in the southwestern states, specifically Arizona. The Indian people were considered possible pickers because of their long experience with cotton. But it was concluded at a meeting in Casa Grande, Arizona, in June 1948 that the Indians would not be suitable because "the Indian does not like heat, therefore, central Arizona, in his mind, is not a choice field for summer labor. Indians prefer to migrate north, during that season, in search of cooler climates. Since his labor is much sought after, he may, almost, make any choice which is suitable to his mood." One notices, immediately, the stereotyping in this description. It was also suggested that the Indians from the northern part of the state, such as the Navajo, would come to central Arizona only when it was getting cold there. In one of the most racist and biased statements reported from this same meeting it was suggested:

> There are many times when Indians will want to go home to attend tribal dances and celebrations or to plant and harvest their crops. Some Indians

> have friends taking care of their crops, but most of them will want to go back home. The most successful method of holding the Indian to his job has been to have him sign a contract. Usually, however, the employer will find that, regardless of the time of the year when an Indian's contract expires, he goes home.
>
> Another interesting hint to farm employers is that the buck will work with the squaw in agricultural labor. Therefore, women can be used alongside men in the fields.[6]

The words *buck* and *squaw* leap off the page. Even in 1948, the dominant culture's attitude toward Native peoples was alarming and had not changed in spite of the whites' proximity to cotton farms and farm laborers. Indians did pick alongside Mexicans and blacks, and Harlan Padfield and William E. Martin's study of work crews in the late 1950s and early 1960s shows Indians as pickers, although not in as great numbers as blacks or Mexicans. Whenever they appeared on work crews, it was as a family or in the company of at least one relative, a husband, wife, sister, brother, father, mother, son, or daughter.

There is no definitive information on when and how black migrants from Arkansas, Oklahoma, Kansas, Missouri, and the South began moving West. It is obvious, however, that the ills that made the Great Migration possible and brought about the Depression of the 1930s affected African Americans in a devastating manner because of a pervasive and historic racism. By the late 1940s and early 1950s, "every year commencing in late August and ending in mid-December, a wave of migratory workers originating in Texas, Oklahoma, Arkansas, and other neighboring states descends into the cotton fields. . . . At its peak in October and November, 2,000 cotton pickers arrive in Arizona every week."[7]

Most lacked skills, education, and financial resources, were of the wrong race and the wrong class, and had the taint and residue of slavery still hanging over their heads. It was no wonder that a father, an uncle, a mother, a son who had struck out on the road West earlier would get word back, somehow, to send another one along. The old idea of "each one help one" as a philosophy made this exodus unique. These migrants also may have heard of or seen the same flyers that the landless whites saw. All of those flyers and posters in the 1930s promising "warm dry winters," "heavy picking," "free tents," "clean cabins" would have been great enticements to come and make "good money" picking cotton. There are many personal tales in these narratives about this part of the experience, and they provide the authentic documentation of this journey West.

Courtesy Arizona Historical Society Library.

Shipping cotton.

Black people never had a love affair with cotton! Despite the hundreds of years of back breakin', shoulder hurtin', knee scrapin', hand bruisin', flesh poppin', chest wheezin', and eye tearin', these cotton pickin' people were never the recipients of the fortunes made from this industry. On the southern plantations and midwestern and western farms, millions of dollars were made, and millions of black lives were lost in the great greed for money. Black people served as free or cheap labor from 1619, when the first slaves landed, to the late 1950s, when automation was introduced in cotton farming. Black Americans have had a lifelong connection to this plant. These "slave-driven" people have contributed greatly to the entire American and foreign economy without receiving even token credit for any of it. So all those cotton songs were not happy songs; they were incantations to alleviate monotony, pain, and the mulelike labor that African American men, women, and children experienced throughout their lives. Also, the various poems and songs written about cotton come from a close, intimate, and uneasy relationship with the plant.

Cotton is not romantic either! Cotton is a rather unfriendly plant, and there is hardly an African American over the age of fifty with southern roots who has not "picked some" or "chopped some." Its stalk is hard to the touch, its bolls stubborn, its pods sharp. But hundreds of pounds could be picked in one day by one person, and more if an entire family picked together. One man said that he could pick a thousand pounds in one day, one woman, "five hundred some." One can only imagine the state of one's back or hands or knees after doing this work for thirty years or more!

Cotton is one of the oldest nonfood crops in the world. Its origin is deemed a mystery in some sources; in others it has been traced back more than four thousand years. It has been located in India, Central America, and South America at its earliest stages, before moving from there to Africa and Eurasia. Padfield and Martin suggest, "Cotton as an Industry is older than either citrus or lettuce. Its origin extends back into pre-Columbian antiquity. There are references to its cultivation in the colonies as early as 1621."[8] In almost all of these countries, laboring with cotton appears to have been relegated to the underclass and to people of color. In the United States the cotton industry was supported by slave labor for more than 150 years. In a 1994 issue of *National Geographic,* an in-depth article entitled "Cotton, King of Fibers" offers many images of cotton's various uses and functions in the economy, yet almost all of them show people of color, such as an East Indian carrying a large basket of cotton to a gin, a Nigerian woman using red cotton as mourning garb, a Mexican costumed in multicolored and embroidered cotton, a Hindu worshiper in India, a young boy in Togo, a black Mississippian driving a mechanical picker, and in Eloy, a black Arizona couple standing in the fields where they once picked. These images also show the close and celebratory connection that these communities have had with cotton at weddings, funerals, religious services, and work. No matter the place, some people do love cotton best. The popular cotton industry commercial sung by Aaron Neville, which says "The feel of cotton, the fabric of our lives," does have some validity. One woman in Accra, West Africa, said, "There are only two important things in life: food and cloth."[9]

The story of cotton in the United States is one of prosperity and affluence. Arizona, for years, has been one of the biggest producers and exporters of cotton. By 1859, "America exported 3.5 million bales or 1.75 billion pounds—or 80% of its crop that year. England was the best customer by far. Mill and plantation owners had become fabulously wealthy." The essay goes on to state that the prosperity of the few depended on the wretchedness of many. As cotton flourished, so did the slavery that supported it. By 1861, when the first shots of the Civil War were fired, some 2.5 million blacks were enslaved on cotton plantations. Solomon Northrop, a slave who labored on a Louisiana plantation in the 1840s, says, "It is an offense invariably followed by a flogging, to be found at the quarters after daybreak. Then the fears and labors of another day begin, and until its close, there is no such thing as rest."[10] Records show that wherever people of any color have worked with cotton, they were subjected to harshness, poor conditions, little or no pay, and, indeed, slavelike conditions.

The history of cotton is long. In the southwestern United States, the Hohokam farmers may have grown cotton in Central Arizona as early as A.D. 500, but the first modern crop was produced in Tempe in 1873. By the late 1940s,

Arizona farmers were planting in excess of six hundred thousand acres of cotton. The most common types of cotton grown in the southwestern United States are pima, which is long staple, and upland, or short staple.[11] Egyptian cotton was introduced in 1901 but was not successful because Arizona lacked the humidity needed for it to do well.

The various Indian tribes of the Southwest, such as the Pima, Hopi, Papago, Yaqui, and Navajo, had cotton farms and for years conducted all of the labor associated with cotton by themselves. They no longer own such large farms, but a few small ones do exist on the Gila River Reservation. But when the boll weevil found its way up from Mexico in the 1890s, the Indians suffered a great loss to their autonomous status as independent farmers. This was followed by the white man's oppressions, the taking of their lands, and their banishment to reservations with land too poor to plant and little or no resources or spirit to even try. The subsequent use of pesticides by white farmers, which prevented the loss of thousands of acres of cotton to the boll weevil, was not an alternative for these Native peoples.

Unlike the thousands of blacks who came looking and hoping for work, the Obed C. Fields family came to the Casa Grande Valley in the early 1930s to buy land and farm its own cotton. Mr. Fields had first visited California to evaluate its possibilities, but he was not enamored by what he saw and returned to Arizona. In the "Desert Beach" area near Coolidge, the Fieldses bought 180 acres and began their own farm. It was worked by members of the family, and only rarely did they hire additional hands to help. Besides cotton, they grew alfalfa, potatoes, watermelon, and cantaloupes. Although it was a prosperous business, O. C. Fields cited very few incidents of envy or racism by whites. Today, his sons still farm about seventy acres of cotton, and old farm equipment can be seen around the property. The Fieldses' experience is noteworthy in African American history, migrant life, and life with cotton. They embodied the metaphor "forty acres and a mule" to integrate industry and capitalism with Booker T. Washington's philosophy of "workin' with your hands so as to pull yourself up."

The work-recruiting campaign, however, was very effective, for about 60 percent of these migrant peoples came directly to Arizona without stopping to work along the way. The flyers were compelling, and a number of blacks became labor contractors as a result of that 1930s movement west. That number grew in the mid-1940s and the 1950s. Perhaps they, too, were continuing the tradition of "brother helping brother." Black contractors were both male and female. The most remembered ones in the Randolph, Eloy, and Casa Grande area are Beulah Moore, from Hope, Arkansas, who had a small operation with one truck; Mary Wilson, who came in 1939 from Fairfield, Texas; and Arkansan Cleveland Jordan, who had a larger operation with four trucks, which he used to transport migrants and food. If there were four trucks, three would have workers and one would carry food. One must remember that the workers had very little in the way of material goods or money to bring, so often they were beholden to the contractors for food, shelter, and money.

The following was said about Mary Wilson: "She was an independent contractor in the cotton industry. She hired cotton laborers and housed them in a hotel which she and her family built expressly for such purposes. The revenue from this business was used to purchase property throughout Pinal and Maricopa Counties. She also fed the cotton laborers meals she prepared from a converted school bus, creating a catering business. Mary Wilson overcame major obstacles in business during that era, dealing in real estate, cotton farming, catering, and assisting laborers in making the transition to permanent life in Eloy."[12] The contracting class became the new black entrepreneurs in the Casa Grande Valley. The Jordan family also had a food business to feed their laborers. Contractors were respected and became the needed voices in these new communities. Often contractors would return to the small towns from which they had migrated—Hope, Arkansas; Hugo and Okmulgee, Oklahoma; Waco and Paris, Texas; and hundreds of other towns in other states. It was not unusual, therefore, to find large pockets of people from the same town working together by the 1940s. Records show that 54 percent of the black pickers were from Oklahoma, 17 percent were from Texas, 14 percent from Arkansas, and 15 percent from Missouri and other states. They formed family crews for picking similar to those of the whites, Mexicans, and Indians. An all-black crew might consist of a mother, son, and brother-in-law; a mother, daughter, and son; or an aunt or uncle and a nephew. The largest "all-Negro" crew recorded in the peak years of the 1950s consisted of thirty people.[13]

There were some crews that had no Negroes, just Mexicans and Indians. Whenever only one Negro was present on a crew, it was most often a single male. What seems obvious here is that since the Negro crew leader contracted with the white farmer to provide a desired number of workers, the workers he used would naturally stay together. In many instances the crew leader or contractor was also a picker or chopper. The cotton contractor's job was to make sure that the cotton hands got to the fields, often in trucks he provided.

By 1948, of the seven thousand migrants in Pinal County, 50 percent were "Negro," 40 percent were Hispanic, and 10 percent were Indian. By December of that year, the total number of migrants had declined to 4,650. However, by then 68 percent of the cotton, or 168,000 bales, was ginned.[14] Traditionally, after

Children with cotton sacks, and picking potatoes in California during the off-season from cotton.

the cotton was picked some families or crews moved on to California to harvest fruits, vegetables, and potatoes. The greatest mobility was among white pickers and Mexicans, who traditionally had more opportunities. School records of migrant children, by race, show that Negro pickers with children more often tended to stay within the region, finding work from other sources. Some black women did domestic work, while some men did day labor. Sometimes one or two members of a family, such as a father and son, or uncle and nephew, would go to California, quite often leaving the women of the family to carry on the best they knew how, caring for their sometimes young and large families.

As discussed earlier, Native Americans, especially the Navajos, were mostly hired at picking time—September into October. By 1952, the number of hand pickers hired in Arizona was between 48,500 and 50,000, with 70 percent

Courtesy National Archives

Toting cotton sacks.

of them Negroes. Sometimes 2,000 people would arrive on a weekend during picking time in the early fall. In addition, "braceros," who were legal temporary help, came from Mexico.[15] Some newspaper articles indicate, however, that only one-third would get work; thus the migrant population was in constant flux. Despite this constant movement, family stability was good because families picked together and pooled their money for the day when they could purchase a home, goods, and a car, or move on. The Pinal County towns of Casa Grande, Eloy, Florence, Coolidge, and Randolph had a great swell in their black population during those years.

Carrying cotton sacks of varying sizes—six, eight, or ten feet long—made of canvas cloth, some with sturdy leather bottoms and a leather strap anchored over the shoulder, hands freed, they walked or crawled between the furrows, often picking very quickly two rows at a time. Some would have knee pads to ease the uncomfortable contact of flesh with earth. Dan Woods gives the following description of a picking scene in Arizona:

> During the "boom years" of hand picking in the '40s and '50s, a big field full of pickers was something to behold. There were blacks, Anglos, Mexican-Americans, and Indians by the hundreds. They all seemed to pick in their little family groups. The sloe-eyed Papagos and Pimas would talk softly among themselves. Sometimes they even seemed to whisper.

> The Mexicans rattled their fast Spanish, the most beautiful language in the world. Seems as though they never run down. And the blacks were great to visit among themselves and to sing and laugh. They sang a kind of calypso type song of their work and their life, sometimes very happily and sometimes the saddest song on earth. Many of the old-time Negro spiritual songs written by Stephen Foster were born in the plantation cotton fields. The black slaves lived them and sang them. To name a few: Massas in the Cold Cold Ground, Old Black Joe, Swing Low, Sweet Chariot, and many more.
>
> The white migrant people seemed to talk of big things and big deals they had been in. Possibly it took their minds off of their present situations.[16]

This description is not only romanticized but is full of racial stereotyping. Woods's essay shows no understanding of the endurance strategies being employed by the overworked and the oppressed.

A good picker could harvest four hundred pounds or more daily, and pickers were paid $3.00 per hundred pounds. If a family picked and put its sacks together, the total number of pounds would be considerable. In some cases a crew leader would contract for more, perhaps $3.35 to $3.50 per hundred pounds. Some boast of doing a thousand pounds daily and making "good money" weekly. There was always a foreman or contractor or farm owner to make sure that the cotton was clean. It was said that the Indians were "very slow and careful . . . [and] picked the cleanest cotton of anyone in the field."[17]

Some black women could pick more than their husbands, and often there were picking contests. There are also stories of pickers who would put rocks in their sacks to increase the weight, but this was usually discovered by the time the cotton got to the gins. Dan Woods reported, "Some people picked for a living, some picked for money to move on, and some picked just for another jug of Dago Red."[18]

Black migrants overwhelmingly did not see this backbreakin' work as denigrating but considered it noble and honest. It solidified families, and many broke out of the cycle of migrancy. Eloy's Art Malone, who became a professional football player for the Philadelphia Eagles and the Atlanta Falcons, says with pride, "It was the only way we could survive as a family. I wish a lot of kids today could go through what we went through. We never left a field because of the heat."[19]

In July 1948, an article appeared in the *Coolidge Examiner* that was an omen of the beginning of the end of hand picking, universally. Entitled "Cotton Picking Machines to Be Tried in Eloy area," it boasted: "One operator can harvest a bale of cotton in an hour and fifteen minutes at a cost of

Aaron Reece, in the trolley used to transport cotton to the train for shipping

$10.00 per bale. . . . One machine will replace 50 or 60 cotton pickers. . . . The most important consideration is to select, prepare, plant, and cultivate with mechanical picking in mind."

Since hand-picked cotton was much cleaner, some growers hesitated to use machines even for the second picking. The ginners opposed machine-picked cotton because it required them to install better cleaning equipment. The controversy continued, but by the mid-1950s and early 1960s, thousands of seasonal pickers were put out of cotton work completely. By 1955, "they broke over the dam of Resistance." At the peak of cotton picking in Pinal Country, pickers were paid close to $2 million in Coolidge, Eloy, and Casa Grande combined, and a total of about $6 million in Pinal County.[20] That money would no longer be there for the newly stranded migrants, some of whom had put down roots, bought land, begun raising families, and established community ties. They couldn't or didn't return "home," so hundreds stayed.

Some black migrants learned how to use the new machinery, but most of the "skilled" jobs went to whites or Mexicans. In some cases, choppin' provided temporary employment. Some did travel to the California's fertile valleys of Fresno, Merced, Bakersfield. Some went to cities as far away as Portland, Oregon, and Los Angeles, California, or to closer ones like Phoenix or Tucson. The breakup of families began at this time as people dispersed to look for work or other opportunities, the same purposes that brought them to Arizona in the first place. These "free" laborers were, again, bonded, this time to the machine age and American white capitalism. Some returned to the valley towns of Eloy, Randolph, Coolidge; some sent for their relatives to join them elsewhere. People were obviously lost for a while, but the human will to survive pushed them in other directions. One former picker who became an itinerant worker said, "It was hard, but we were better off back then."

There are a number of events that continue to focus on the importance of cotton in the Casa Grande region, such as the regular meetings of the Cotton Wives Club, the Cotton Princess Pageant, historical museum exhibits, and the Coolidge Cotton Festival, which used to be held in late fall after the harvests but is now held in February to take advantage of the money hundreds of "snowbirds" bring to Arizona. None of these draws on the surrounding black communities that contributed vastly to the cotton industry's success.

Obed C. Fields

I had heard and read of the Fieldses of Randolph and Coolidge several times. In every instance, it was always in connection with cotton, black independent farming, established old families, and distinguished citizens. The story of a black Oklahoman who came not to pick cotton but to buy land to farm, and who was and still is successful, was worthy of a chapter in Arizona history by itself. I thought of the Fieldses as celebrities, so I was rather timid in approaching them. One more interviewer in their lives would certainly be turned away! Additionally, the Fieldses—Edna and Obed—were now both in their nineties; in June 1992 O. C. was ninety-two, his wife almost ninety-four. What about their memories? Would the children or grandchildren run me off? Not so. After calling and setting up an appointment to meet them, I had a wonderful visit in their nice and spacious home near Coolidge.

After some misdirected driving along La Palma Road from Randolph Road, I found the correct address. Obed Fields greeted me and escorted me to a chair in their parlor, not the TV room. There were several comfortable chairs and a sofa. Their house was clean, tidy, and cheerful, and they appear to have lived much better than their Randolph neighbors. Photographs of Mr. and Mrs. Fields were on the walls, as well as photos of Mrs. Fields's parents.

Mrs. Fields is a neat, quiet woman who listens well and speaks only when appropriate to add to or correct something said by Mr. Fields. But one can feel the bond, respect, and partnership they share after so many years of life together. She wore a bandanna on her head and a nice clean housedress with a small, matching apron. It appears she dresses like this every day. She has a certain sparkle when she speaks about rearing children and how tough it has gotten to do so. Mr. Fields, she says, is "still going out to pick a few weeds and watch the cotton grow." They now have seventy acres on one side of the road, and forty more that they rent for farming or sharecropping. Their sons, Joel, Romeo, Samuel, and Calvin, do all the work connected with their cotton crops. Joel is seventy years old and lives behind his parents' home. During the interview, Mrs. Fields said very little, but she hung on to every word for more than two hours. The narrative has only a few sentences from her, which are given in italics.

Both Mr. and Mrs. Fields seemed to be in good health, except that Mr. Fields's vision was not very good, although one would never know it, and I have watched him read and write without glasses on many occasions since that initial visit. As our two hours together were ending, Mrs. Fields said, "Oh, my goodness, we've been sitting here all this time and I didn't offer you a drink." She got up promptly and brought a clear glass of iced water and a Pepsi; "Whatever you prefer," she said. Her demure manner was apologetic as if to say, "I have manners and training but slipped—my, oh my." It was a rare moment with her that remains vivid for me.

Mr. Fields consented to having his picture taken, but Mrs. Fields hid her face and would not let me take any photos of her. I assented to her wishes and thought it best not to push. She did, however, invite me to attend their church, the Trinity Baptist Church in Coolidge. I promised to do so. It was a memorable afternoon spent with two first-class, gracious elders. Their reputation of welcoming people to their home in the old days proved true. Meeting them was one of the highlights of this project.

Edna Fields, born on November 21, 1898, in Marlin, Texas, to Peter and LouAnna Williams, the oldest of their fourteen children, died on March 2, 1994. She was almost ninety-six years old and had been married to Obed C. Fields for seventy-two years. Mr. Fields, born in Falls County, Texas, celebrated his ninety-seventh birthday on February 13, 1997. He continues to oversee his cotton acres and his sons, irrigate, and pull weeds.

Well, I'm O. C. Fields, and I'm ninety-two. My wife was born in 1898, and me, 1900, February 13. Yeah, I seen a lot of years come. I was born in Texas, reared up in Oklahoma. Born in Texas, part they call Falls County, Texas, that would be the south part of Texas. We came here in 1932 from Okmulgee, Oklahoma. Came here with my father, and two dollars in my pocket. My father had a little difficulty back there.

My father was a farmer, in Texas and Oklahoma. Moldis Fields was my father; Ella Fields was my mother. Yes, they was farmers, Grandmama and them. And then I came to Arizona—left Oklahoma for, well, I guess just different things, you know, just traveling around. See, I'd been in California, and me and my father-in-law stayed out there quite a while. And we didn't do so well then. We finally went back to Oklahoma, and then I came back to Arizona. Yes. I was a pretty young man then, something like in the thirties or something like that. Of course that's been about sixty years ago. [Laughter] Came out here because I was interested in the open life. I wasn't interested in the cities. California interested me, it was fine, but 'cause of the crop and agriculture part I didn't do too well there. El Centro I believe it was.

Many people came to pick cotton back then as migratory workers, but my daddy wanted to try farming. When we started out, we worked for a rancher, the Woodruffs. Yes, they raised racehorses. It was two sisters that owned the ranch, white sisters, but it was operated by one man, Major Woodruff. So, I worked for him. That was right out from Coolidge. And then, on from there, I evidently was with the Woodruffs something like two or three years. And then I worked with the Gammage ranch, Melvin, Melvin Gammage. I worked with him for eleven years. I was doing farmwork, irrigating and whatnot, just regular farmwork. I think that's some of the same Gammages with a lot of money and parts of Arizona City University [now Arizona State University] named after them. Yeah, I worked for them eleven years. Well, the year we started farming cotton was . . . that's been way back, I don't recall now. I came out here in 1932, bought the land after working for Woodruff and Gammage, so maybe 1946, somethin' like that.

When we first started workin' our land, we didn't have the modern equipment, we just had the old-type equipment. But finally we went on into later equipment. We farmed that eighty acres, and of course they didn't have no water, had a well—yeah. Just worked it with our own family. It was a full-time job, yes. Our children, they were, well, our son was grown, and they were young men. My oldest son, Joel, is seventy now, he lives behind us here. Our land goes back this way [pointing]. We had eight children, three girls and five boys. Yes, eight, but we lost one. *(Mrs. Fields: It was about three or four days old.)* The others is all living. Some of our own children didn't farm, but we did alright. So on like that.

And since we was interested in buying our own, we bought our eighty acres, and then we bought this south forty. Really 120 acres, yes. I don't really know for sure now, but evidently it was fairly cheap, see, because this here, I think we paid sixteen or seventeen hundred dollars for that eighty. I think that was it, somethin' like that. We didn't go through the bank. Cash, yes, cash. We bought up that eighty, and then we bought this south forty, I think it was two acres, a forty-foot strip off of this forty. It was just vacant, just desert land. Bought it from Ernie Hall. We just interested in farmin', you know. I liked to grow cotton. This was just open land here, but later we drilled a well on this place.

Our land went down to the corner, goin' this way [pointing]. See, this Randolph Road, on the east side just as you turn the corner down there . . . see this, La Palma Road plays out right down there. So we added on that eighty that went right back east there, and then this forty. Yes. And then we drilled a well on this place. It was Carliss drilled the well for us. So that gave us water with a well on each place. Then you had to have your pumps. We used to use the pumps with motors, but, finally, we went to electricity.

We would have just one crop a year. Spring planting, harvest in the fall. We just picked by hand at that time, but now you use machines. Yes. So at first it was just hand pickin'. Once we got it picked, it went to the gin, ginned out into bales and bags. We sold it and went to the different storehouses and whatnot. We just always did our own work when we was harvestin' by hand, didn't hire anybody, just the family. *(Mrs. Fields: Sometimes, you know, ladies would come by and wanted to pick. And then the people out of Mexico, those Spanish people, they was lookin' for work. The farmers wasn't supposed to hire them, but they hired them cheap, and we wouldn't deprive them of a job, because they were cheap, you know, they'd do it real cheap.)* Yeah. We worked at that, evidently, three years—let's see, seven or eight years—before we got the different equipment. Electric pumps and electric motors, you know.

Edna Fields.

Sometimes in the spring on top of the hill, between rows of cotton, we would plant watermelon and alfalfa. So, we did pick acres and acres of watermelon too. We was never unemployed even after the 1950s and '60s when the machine came in. My wife worked

The Eleven Mile Corner cotton gin.

as a maid on different farms sometimes. She had a third-grade education but made just about all the furniture in this house. Good with her hands. Yes. *(Mrs. Fields: I used to have lunch wagons to sell food too. Yeah.)* The one-row machine and two-row machine put a lot of people out of work. Machines, machines, everything to the machines! My son still works on the old ones back there. You can see him back, behind the house; he lives next door. My other son, Romeo, lives just a few yards the other way. Yes. We's still all together here. Romeo works for the Coolidge School District in transportation. Another grandson, a teacher, intermarried with white.

I met with prejudice few times. In the 1930s, I was denied full pay. One rancher only paid fifty dollars a month. I went to cash a check in Coolidge and a white man said, "That's too much money for a nigger." I remember also the Woodruffs' son saying, "What's dis nigger doin' here!" I remember those, yes. We presently have seventy acres of cotton in production; my sons do all the work, the one that works in Coolidge, you know, whenever he get a break he help with the irrigatin' and like that.

Farmin' got to be more difficult lately. You got to produce so much, plus expense to break even. Not profitable anymore. No. We also have more problems with competition from the other farmers; not everyone is friendly these days. You got to be almost a lawyer to farm now. Yes.

Nick White

Pastor Nick White claims Oklahoma as home, even though he, like Mr. Fields, was born in Texas. He was one of the first persons I met in Randolph. He and a grandson were working on his church, which stands at the corner of King and Bell Streets. He is the pastor of the Church of God in Christ, a position he inherited from his father. He mentions in his narrative how he got involved in the church, other businesses, and the Randolph community over the course of fifty years.

We had two long interviews quite a few months apart. The second took place in his church on a scorching hot day. He decided it was cooler than his house. I agreed. Not only did Mr. Nick relate his life in his cheerful way with a broad smile and eyes that light up when he speaks, but he was one of the most patient men I had ever met. His memory for detail was keen, and he seemed to like the remembering. He would also repeat phrases with a look that reflected a happy time and longings.

He liked to repeat anecdotes, and one of his favorites was his inability to pick 500 pounds of cotton in one day. He'd get to 492 but could not get the other eight pounds, promising to "get it tomorrow," but tomorrow never came. When he'd repeat this story, he would throw his head back and laugh. He also teased his wife for always outdoing him in the field. Another story involved their lack of money and suffering in Oklahoma, indicating that they just always "worked together for the crop man." The trip west to Arizona, by car and trailer, is still vivid for him. Nick and his wife took me east one day to the old Route 60 that comes out of Globe and "the mountains" to demonstrate how difficult and narrow the roads were. He also helped me to reconstruct the town of Randolph in the old days—the 1940s, '50s, and '60s, when it was flourishing.

This smiling, warm, and patient man is also a community observer and dreamer, a compassionate person, who agonizes and worries about those less fortunate than he, the younger generation, and the future of towns like Coolidge and Randolph. He says: "They had the stores, the department stores . . . grocery stores . . . service stations. A bar, a tavern, a little of everything. But machines came in and the people didn't come in from Texas and Oklahoma and them places anymore. . . . So one store after another closed up. . . . People started leaving. . . . You see in Coolidge vacant spots like here in Randolph. The buildings been taken down. Everything. Even the spark, even the joy in the people's eyes."

Pastor White also has a profound and spiritual philosophy about helping people. He says, "I don't crack down on people, I just give 'em my last, takin' my chance on some of 'em. . . . So that way you live and you let live. Give and it shall be given to you. Good measures, pressed down, shaken together. Good measures. Runnin' over. . . . This is a true sayin'. . . . I don't know what the future of Randolph is. It looks bad, but I guess it could be worse, you know."

Besides the two extended interviews, we have talked by phone and on other short, informal visits, and he always seemed so happy to see me. He told me recently, "You have made many a trip here, haven't you. God has his reasons." The White family has many relatives in Randolph—sons, daughters, grandchildren, great-grandchildren. Their door is always open because theirs is a house in constant motion.

Nick White's memories of "workin' in cotton" are not filled with regret or remorse. As he compares working in Oklahoma with working in Arizona, he says: "In Oklahoma we didn't make no money. We just worked together for the crop man." He also adds, "You know, I think we did better in them days than now; at least you knew where your kids were. The family would all be together—pickin'."

I was born in Willow Springs, Texas, in 1926. We moved to Oklahoma in 1930. In 1943 we moved from Oklahoma to Arizona, to Randolph. Had quite a trip comin' to Arizona. I came to Arizona ridin' in a four-wheel trailer. I wasn't ridin' in the car. I was ridin' in the trailer behind the car. In the car and trailer was seventeen people. We left Oklahoma, goin' to Oklahoma City on a Sunday evening at one o'clock. By daylight, the next morning, we got to Oklahoma City. We drove one hundred miles.

The car we was drivin' was a '35 Plymouth, and every time we started, it would cut out and quit. We spent two nights and days in Oklahoma. Two nights and days in Texas, one night in New Mexico, then to Arizona. We got to Coolidge, I guess, sometime before sundown on Friday. Started on a Sunday, and got to Arizona Friday evening by sundown. I was sixteen when I came.

We got to Mr. Fields's house because my uncle and Mr. Fields were neighbors. They were livin' right close together. They had bought eighty acres together. Each one had forty acres of land. They was our folks out here. So we stayed there and we start pickin' cotton. And it was seventeen in the car and truck. One of them was my Uncle Silas White. Silas White, Mudrow Pete, Grace Pete, Walter White, Ella Bea White, Nora White, Thomas Johnson, John Davis, Lonzo Davis, Jonna Lee Davis and Nella Davis and Ezekiel Boston, Thelma Boston, Adella Boston, Doris Boston, and Idella Boston. All relatives out of Arkansas. So the whole total was seventeen.

Top speed from Oklahoma to Arizona was almost twenty-six miles an hour. Every hill in Texas, everywhere we were tryin' to get up a hill, he would ride the brakes down the hill and we'd be gettin' a little speed up, he would wait to get to another, to get anything that would start it. Every so often, it would break down. We had to stop in Oklahoma. Put a fuel pump on it. Havin' trouble with the car. Soon as the evenin' sun dropped, we'd stop. You only travel at daytime. Just like a wagon train.

But it was exciting, though. And we made it to Arizona, and we got set up here. Went to the field and picked cotton that Sunday and Monday. And me and my mother and my brother, we made fifty-four dollars all together pickin' cotton the first week. The next week we made identically the same thing. And that was pretty good. In Oklahoma we didn't make no money. We just worked together for the crop man. Every little money you made was cash.

My mother, me, and my younger brother, Elbert, when we first came here and we picked cotton, that was all of the family that was here. Yeah, just us three. My daddy, he yet in Oklahoma. He had to farm, he had to finish gatherin' his crop in Oklahoma. He drove his own car later on. About Christmastime, he came on. After he gathered his new crop in Oklahoma, sell what he had, what he could, and came West and came to the state. Him, another man, and my two little sisters, they came over. And so we got in good shape then because we didn't have no money in Oklahoma. Just could pay back two hundred and some dollars he owe. So Daddy got no money, you know.

Somewhere in September we came. Cotton pickin' just gettin' going, gettin' started. September, first of October. Anyway, we came in, went right on into the field. We didn't have no place to live, so we stayed over here by the Fields a few days. Then we moved to a little ol' camp over there near Coolidge. We didn't have no beds, we didn't have enough room, so we had a pile of cotton in that little house. They had another house where the womenfolks stayed, and the men stayed in that other house, and we just sleep right on the pile of cotton. All of us lay around just like lumber layin' up there.

Up the next morning go pick your cotton, come back and clean up, go to bed on that pile of cotton. We did that for two or three weeks or better. My Uncle James got a tent and set it up, and he come move me and my mother and my brother over to his place so we could harvest the fields. And that's where we was when my daddy came. My uncle didn't let us stay over there. He got us out from the cotton fields. And we stayed there and we stayed on his place there pickin' cotton for this man, Mr. Gamez. And my uncle has forty acres of land hadn't been cleaned up. Sometime in the fall, when the dew was heavy and we couldn't go to the field early, we'd go out and help him clean up his farm. We stayed there, and he built a tent for me and Mama. We had it pretty nice there. Then others got different places and they moved off. We didn't have to go way out there on the camp, we rode the truck out to the field and back.

Saturdays, sometime you pickin' cotton, you come out of the field and stop by the store. But when we workin' in the field, my mama she'd cook and we ate like kings! She'd count enough food for the field for us, just like a picnic. We eating two or three times a day, me and my brother. She even fix breakfast before we leave, and she had so much stuff for us to eat anytime we got home! She'd fix it early in the morning and when she got off of work. She didn't pick no awful lot, but she picked. My brother picked the most. I couldn't pick as much as he did, but I'd pick, then I would carry my mother's sack. We made it pretty good. We didn't make nothin' in Oklahoma, nothin'. But by Christmastime we saved up about six hundred and some dollars. We was in good shape.

Let's see, pima cotton was three dollars a hundred. You couldn't pick much pima because of the type of cotton it was. If you could pick a hundred, you really done good. And I couldn't pick that, but I could pick some cotton! Mostly, I picked the cotton and I carried my mother's sack. My brother, he was a good cotton picker. He made the most money, but I made up by makin' him carry my mother's sack. So we saved a little money. When I did fifty-four dollars, my mother would save the money right off the bat. And another time after Dad got here we just saved up six hundred and some dollars. Then we picked cotton until March. We was hand pickin'. We picked cotton all the way to March. After that, after the cotton was over, I went to California and worked in defense. That was durin' World War II.

Yeah, we picked cotton from September all the way to March. Just took that long to pick it. There's no machines. Start pickin' around in August, September. Cotton stay on there; begin to open, just begin to open. And if you hand pickin' you pick it about twice. Then when that frost come, you go back and you pick it again. Sometimes we had a hundred people. Maybe that field might be eighty acres. They got so many fields.

Pick this field out, sometime it take you two or three days to pick one row or half row of cotton. It would be that much cotton on a row. You'd get a row and pick all that day. Pick the next day, sometimes you pick the third day to get that one row out. And that was a lot of cotton to go off the row. Yeah, you pick it, carry it to the trailer, and do it all. Some of 'em are twelve-foot sacks, some are ten-foot sacks, some nine-foot; eight-foot sacks, then a three-foot sack. We most always pick the twelve-foot sack. And you pack it tight as you can. So that was hard work, strainin' work, but people, they didn't worry about work, just the little money they was makin'. They didn't make that much, but it kept it in circulation. They made it, they spent it, everybody had a little money.

They picked fast 'til Saturday noon. At Saturday noon they'd get the pay, and they go to town and to shoppin' and whatnot. Saturday noon or Sunday they rest. Monday mornin' get right back out there, about sunup. They hit the field and then they pick all day. I never could pick a whole lot. I never did pick five hundred pounds. No, I picked three hundred some or four hundred some every day, but I never did get to five hundred!

Now, my wife picked five hundred pounds every day. She picked five hundred every day. And then the time I could have picked five hundred, I said, "Well, I'll get it tomorrow!" And the sayin' is, never put off today for tomorrow, what you can do today. 'Cause I had plenty of time to go back out there. When I got 492 pounds, all I had to do was go back out there and pick that other eight pounds. But I said, "I'll do it tomorrow." Tomorrow never got you. So I never did. But we would pick about nine hundred and some pounds of cotton a day. Just us two. We didn't have no children then.

So cotton finally went up, it went to three dollars a hundred. If you pick three hundred pounds, that's nine dollars. Three fifty, that was twelve dollars a day, you go out there and pick it. And your wife pick it and you count that up. And you adjusted yourself accordingly. But you could buy a whole lot with that. I drove a better car then, way better than I do now. So I had a 1953 Dodge.

I bought one; my brother-in-law, he bought two in one year. He bought a V-8 Dodge, he got me to take it over then, that had only eight thousand miles on it. And we went back and bought him a new Oldsmobile. I paid for that car. I was makin' seven dollars a day, six days a week. We paid for the car, I paid for Dad, I bought my mother and dad a bedroom set. I made the payment on that bedroom set and I paid for that car. And all other expenses. I did that makin' seven dollars a day! Yeah. I'd go get my check and I'd go to the store and drop in my seven dollars a month, ten or fifteen dollars over here, and saved some.

When we'd pick, we had fun, you know. And then Mr. John Davis and his daughters were in together pickin' a thousand pounds a day. They had a brother that did six hundred pounds. He had three children. The children picked so much cotton, their daddy would keep the weights. He bought lots of land up there. That was Mr. Davis; he's dead. Most of that land up in that corner there, on this side, and that land up there on that corner with that sign there, that's his land there. And most of this like across the back on this hill is his. John Davis was a black man. We all come together. He was on that wagon, on that truck. You know, he come, it's that four-wheel trailer back there. There were six in the car; the other eleven was in that trailer. And John Davis was in that trailer. He was on that trailer. He wore the seat of his pants out, you know.

But when we came, we got to workin' and they picked cotton and everybody had some money. They didn't make nothin' at cotton hardly, they didn't make nothin'. Back in Oklahoma.

After all the cotton was gone we went to California with my dad. And so from then on we did pretty good until they tried farmin' one day and my dad wouldn't farm. Well, his brothers messed him up. See, my Uncle Sy and my Uncle James and my dad, they went together, they rented a place down here next to this place, right over in this gatherin' here. They farmed it, and sometime we was stayin' places. You think you know how to do somethin' and you don't really know.

My Uncle James, he'd been there in Arizona for years, and he told them all just how to farm the cotton, how to feed the plant, and so forth. And so my dad and Uncle Sy and them was his younger brothers. And they planted the cotton like he told 'em, too deep. So they had to replant it. That throwed the crop late. They made a good crop but it was late turnin' it in. The finance rates gone up, and it didn't make. They didn't go too much in the hole, but they didn't make no money. They didn't make no killing the first year. When my Uncle Sy went to Oklahoma, my Uncle James didn't say nothin'. And so my daddy was broke. And the man wouldn't wait to collect. Now if he went back and farmed it next year they'd have made some money. The mistake was they

impression of success, more stability here than in Grapes — but folks left to be interviewed years later by definition achieved something left my daddy out on the limb. So he was kinda just stuck with the debt. But he worked on the side, and he worked 'til he got about sixty somethin', and he got sick, you know. He died when he was sixty-six years old.

It's a cemetery there on the road between Coolidge and Florence, you pass right by it. If you're goin' into Florence, it's on the left-hand side. As you go over the overpass, about a mile or some after you get to the overpass goin' towards Florence, it's right there. There's a farmhouse that's back there. There's a bunch of oleanders, a house on one side; the oleanders on the other side of the cemetery. I guess the oleanders are on the east side. Used to only bury blacks in the back, but I know some of the blacks started burying the folks up to the front. But most of my uncles are buried over in the back.

When I became eighteen I was drafted into the service, trained at Fort Leavenworth, on to basic technical trainin', and I went to the Philippine Islands. I came back to the States and I married Metris White in 1947. We started in pickin' cotton—it was cotton harvestin' time. It was hot and dry and we picked cotton, chopped cotton, day in and day out. So then we was choppin' cotton, we made it pretty good. At cotton pickin' time there was a space between there for work. Pickin' cotton time we would make enough money to carry us over to the work period again. So we picked lots of cotton—we picked hard.

Me and my wife, we would chop cotton and had about everything that we finally need in the house before we ever married. We worked, chopped cotton and picked cotton. We went out and bought our house out there. We had a bed and icebox, kerosene range, like a little gas range. All we had to do is set it up. It was a long time from the time we married 'til we got the children. But we got 'em, we had a good time. The more children we got, the less we had to bother. The thirty year old, he'd be traveling all his life. The children, they were no trouble, there were just the two girls and two boys.

I ate well, and my wife she sometime would chop cotton. Then that one time we was choppin' cotton for a contractor, the contractor went off with our money, didn't get no pay at all! He took all of it. But it got him in trouble. But he took the money and went off and gambled it all. About eighty some dollars, me and my wife made forty some dollars apiece choppin' cotton for him. But he done gambled it off.

Right now, I don't think you could even find a cotton sack. I went over to Mexico several years back and I seen a Mexican woman and an old Mexican, he was pickin' cotton with a sack. And it was amazin', you know, it was somethin' to look at. Just to think I used to pick, you know, we made our living like that—made good money! They may be still usin' them across the border in Mexico. Maybe in Acadonia or San Luis, or somewhere. They were either in

Mexico or Yuma. In the late 1940s, work began to change, a lot of machinery begin to come in and less cotton pickin', so we went away from cotton pickin', we went to construction work. My wife got a job at the Training Center.

You see the people around here, they had a good time, developin' land then and begin puttin' up the houses here and everybody had their house. All the folks come out here to stay. John Davis had a house up there, had some land over here. This house I got, me and my dad built it, and my dad built one out there, and my dad built several other people's houses. He could even build a chicken coop. He never went to no primary school. And when we would come to Arizona, when the first pickin' was over, he went to Phoenix and joined the carpenter's union. And when he was initiated, he had to go there and read that blueprint. Where he learned it, I don't know. That was his initiation process, readin' that blueprint. They'd take him out, that was the first time I ever went to Phoenix, you know. He said, "First some cotton pickin', come on up here, I'm gonna be initiated." I never went to Phoenix until I rode up with him. He joined the union, was initiated that night. When I got to Phoenix and there were all these trees and all the fruit trees and I said that it was the prettiest place there was. I'd never seen it, you know. In Oklahoma, you didn't see no orange tree, peach tree, or apple tree back there. But orange or grapefruit, I've never seen that 'til the trip to Phoenix. I say, man, this is the place I wanna be. I wanna eat them things. Them trees had the flowers, Japanese flowers on it, and the oranges and grapefruit and all was ripe.

When we got the Migrant Opportunity Program in here, they helped a lot of people to create jobs. In fact we created a lot of jobs for the folks 'cause we had to redo the church, 'cause we just had the barracks. And the barracks we had, it was one of the barracks that the Japanese used during World War II.

The Migrant Program was in here durin' the time that President Kennedy was killed, 'cause I was workin' at Picacho buildin' that freeway. I helped build those freeways. And it was sponsored by the church, so I got to be president of it. A preacher usually sponsors it—and since the church was the sponsor, I got appointed. So I was the president, but just about every time on the Sunday they'd have the meetings I had to read for the church. So I got to be pastor of the church. In '65 I was appointed pastor. And so I had to be at the church on Sundays. So this went on until the year I resigned being president. And it worked, it would grow, and a whole lot of improvements started in the community because finally we got the streets paved and finally got some water in here.

Well, we got the water in a little before we got the programs in here. Because we hadn't had no water until we got the programs. We used to haul water from over Twilight Trail way over here. We used to carry water from this well right across the street over here. And everybody hauled their water. The streets was dusty and dirty. You'd go down the street in the evenin' drivin' fast, the dirt just covered everybody. We can't move. [Laughter] I used to walk from my house; at that time I had a store. I bought the store from my uncle, right over there, second house from the end. That was the store. The second house on the end, and the house this side of it, that was the store. Just straight across the street from my house. Second house was the store. That store and that house on the end belonged to my uncle Simon. So he sold it when he went to California. Later on, even before we got water here, my dad and I moved the store. I put in a service station on the highway right straight behind the store and put in the gas. Then I moved the store into the service station.

And so when I opened the service station, the amazing thing was, the Texaco station is the closest thing to my Chevron station. So I needed some money and needed a cosigner. I went down to the white man, the one that run the Texaco station, and he cosigned for me to borrow money to put in my station. So that's the kinda fella he was. And somebody asked me, "You mean to tell me the Marshall brothers helped you go into competition with him?" I say, "Yeah, he did." He did and paid the money—I opened the business up. When I opened the gas station, the gas was $.3750 for regular. From the end of 1960, to 1979, the price was from thirty cents to forty cents. It was better than pickin' cotton. And I sold a lot of gas. I got tired of it because I didn't have no water, had to haul water to the station, to put water in people's cars, and water for the person to use. That was from the latter part of the '50s into the early '60s. And when I got robbed that day, I had no water. I had tried to set up a system so we can handle water and have a bathroom where the customers come in, but I never did get it completed.

I kept that business three or four years. My brother-in-law had a store, and he went out of business. He had bought a meat counter, meat slicer, and I bought all that equipment. I had a place where it kept the meat cold. Fresh meat. The ice man, he put an icebox in. I sold ice outside. And I sold gas in the front. By that time the cotton pickin' was in a big slump. I picked a little—my wife would pick some. I mostly had to stay at the station. A lot of people was here in Randolph then, and houses and cabins all down the side of there. Even right up against the church. Houses all over there, all up here, some up on the hill, some up here, some back toward over there. Just full of people. Oh, it was better than twice the people that it is here now, which I'd say is about two hundred and some people or better. There was quite a few Mexicans. They mostly stayed at that end and up on that end, to the north, and to the south.

And so, by the time I put the station in here, Myrtle Jordan and her husband married. They used to have a camp over here.

They got married and they bought this place, had eleven black guys. All this land out here was the Scotts'. He sold the place to Myrtle and her husband, Cleveland Jordan. All but one or two lots right out there. He got 'em right there. He lived in Stockton, California. Anyways, they moved to Randolph, opened camp. When the cotton machine would come in and people would be movin' off the camps, we went in to get them regular homes.

They had a little bar here. John Davis had a little ol' cafeteria thing up here, a restaurant. He had a little café up there, and they played music, and they'd dance and all that. His wife belonged to church. The children, they was in and out of church. And so, then after the cotton changed, then they begin this moving out. They gettin' too old to work, the children got grown and all. Then the parents moved to places where they didn't have to do all the hard work. A lot of 'em moved out of town. A lot of people from Missouri, Texas, and Oklahoma. All of 'em didn't go home, some were livin' in Phoenix and other places. But I guess I didn't have sense enough to move. I ended up stayin' right here! Some went on to California. Even the fella Thomas, that came on the trip with us, went straight on to California. He stayed and picked cotton a year. That's pretty good money in his pocket even when all his folks was in California, his brother and such. So he's still livin' there in California, Los Angeles. I saw him over there. I see him every time I go there.

I remember Mr. Babbitt, I think he was a farmer or a schoolteacher. He had the post office in operation when I came to the state. That's where we used to come get our mail in there. He was a nice old man. He was real old all right. But he was nice when we first came. Then he had an old '33 or '34 Chevy truck that he would take the mail up to the railroad track and hang it on the bay, on the little thing they had hangin' there. Then he'd sit there and wait till the train come by sometime. Pick up his bag and his mail, he'd throw it up on the rack. He did that all the time 'til old man Jesse Jordan be doin' it.

There was one store in Randolph and one service station. You go pick cotton and when we'd come out of the field, they'd go to the store and buy their food and drinks. This was in Coolidge. And after that year was over we went to California to work, and I went in the service for about a year. Army. I come back then to live here.

My daddy was a carpenter. My uncle was a carpenter. They was workin' on a buildin', they was just completin' it and it was nice. After a few weeks they have a school in there. But it burned down. So then we didn't have no buildin'. After the war was over, a Japanese prison came in. They put those Japanese in them concentration camps at Florence. When they moved out, they sold some of the barracks. So we bought a barracks from Florence and set it on this old foundation church. We didn't got the money to raise it, so on Saturdays we picked cotton. Saturday we picked the chewin's, Saturday mornin' to Saturday noon. But everybody picked and they gave it to the church.

Yeah, them old men, Mr. Bell, my dad, and Sy. All of them. They picked cotton hard. They picked until they had several hundred pounds of that cotton a day. They do that every, not just one Saturday, *every* Saturday. They did that. And they bought that buildin', bought it clear and my dad then fixed it up so they could have church. So, for forty years they had that wooden church out there. And the only thing left out of that wooden church is this pulpit. We tore all the rest of it down, we couldn't get it off of there so I slid it up there for a pulpit, after we got the buildin' set up.

This part of the pulpit, that's the last of the barracks. The barracks was a kitchen, and it was a dinin' room, and they framed it up and they poured a concrete floor inside on the wood. And they had to take it and pry that concrete, about four inches of concrete, and break it apart and unload it on the truck and get the concrete out of it so we could move that here. So they did that, they moved it here and they set it up and we used it for years. We added on to it. And that's how it was built. No debt. No mortgage loan, nothin'!

I got in church right after I come in after the army. I got in church before I married about February of '47. And so it took all the running out a ya, too much runnin' around I figured. [Laughter] I went to Florence County. I got a license. I didn't waste no time. If I go do it, I go do it. And I got married on a Saturday morning, May the thirty-first, the last day of May. I was in the church for nineteen years. I started preachin'. So I've been in the church forty-seven-some years.

That first church was originally in Coolidge. Later the colored folk begin to move into here [Randolph] and to buy land and to get developed. Then there were more blacks out here than there was in Coolidge. A lot of them around Coolidge. Especially churchgoers. They had a Baptist church on the other side of Coolidge. They built that. The church up there relocated out here.

Our first church building burned down, on this same location, and was rebuilt. See those dark spots there? That's where it was burned. Forty years later I built on it again. Same pattern, I extended it out twenty feet long. And then the same church here, after they had the barracks on the end, wooden church, somebody set it on fire. They burned it up. Set the pulpit on fire. I had a back room and a bathroom in the back, went back and set the curtains on fire. It was wood. Dry wood. Nobody knowed it was on fire. They had pillows on this bench here, it was stacked about four or five high. They set the pillows on fire. The pillows burned right through this wood. I put new board on here. New wood. Pillows burned right through it, and the acid spilled right through on the floor and it all went out. The pulpit was burned up, we had to put a new one in. And the pulpit up here is in there. It burned and scorched it all up. The curtains that caught on fire back in the back, it went out. Wasn't nothin', only God wouldn't suffer the burn. 'Cause we didn't know nothin' about it.

I think it was kinda in the spring when this happened. We was workin' somewhere at the time, 'cause I worked construction. So when we knew anything, it burned out. Somebody come up here for somethin'. Mrs. Smith found smoke in here. Found the pulpit. The ashes didn't even scatter when it fell through there. The pillows just stayed in a neat circle. And the ceiling and floor, the molding here, just burned a round hole right through it. I'll show it to you in a minute. It just burned a hole up here. After I was made a pastor, I had a man to make a new one. This is not the original one. It burned it all up, just scorched all around. All of that burned.

Now, these wooden pews, we got those later. All these chairs. All the chairs been there for years. The wooden ones I got, they were given to me later some years back. Those back here in the choir, they was there. Got them at the same time. The original church, everything burned down. But after we got the next building, that's when we bought the chairs. The wooden building. Remember, my daddy was a carpenter. When they built the first church, another man came down and laid the blocks and built it. And when the church burned down, there was school in here. They had a public school. I think some of the schoolkids did something to the stage like that, but they used it for heat. The teacher stayed there with Clara Bell. She was a teacher here, way before the Migrant Opportunity Program school.

All the black children went to school at Borree Corner. And then before they really integrated them, some of the kids went to the integrated high school. The grammar school kids went to Borree Corner, and then later on they start sendin' them to Kenilworth. And then to Coolidge.

That Borree Corner School was in session when I came to Arizona. They had school over there. They had an old church over there. They had school in that old church. I don't know if it burned down or what, then they moved it down to Borree Corner. There's an old bar there now. I used to go in and visit before I went in the army. Well, it was on that side of the street. The teacher was Miss Miller. And then after that they moved it across the street in another buildin' over there. And they had a man that was white teaching there. I can't think of his name. And then the school came to Randolph. Then it burned.

It's really not many of them old buildin's here, a lot of them are burned down. My father's house burned down. The Jesse Jordans' house burned down. My sister's house burned down. Borree Corner School burned down. The church burned down. And then five houses, Mrs. Bell had, after Migrant Opportunity come in, she had moved the house off her ranch. That house sittin' there, it burned down. Cindy's house burned down. Six houses burned down at one time. Think some little boys just hid in that tall grass like out there. Sit there in the grass, the wind was blowin'. Burned all them houses down at one time. Everything together. The first house, it was empty. The rest of them folks had everything burned up. The fire station come out there. Then the Marshall boys called the county trucks, and they brought water from the Training Center up here. The only reason why daddy's house over there didn't burn down was they was sittin' there and skeeted water on it. And yet it was smokin' like this.

And it was one of the biggest fires I've ever seen. Six houses at one time. Wind just rolled that heat. I had some hogs out there in the pen right in the back of the yard and the heat was so bad I tried to hold out and run 'em out and they got excited and turn right around and went back and got in the pen. Any other time they'd run all over the place. All that heat. But they survived.

Cindy and I fought that fire. I had my hose, and the fire begin to spread toward the church again, and I got it turned away from here till it got stopped. But nobody got hurt. Even one of the men that had a house that burned down, he wasn't even at home. Some of 'em had too many things in the backyard. But you know how we do sometime.

I seen Randolph grow up, and I seen it shrink down, you know. I seen it when it was in a boom, and then I seen it like now. Like the time when the people was picking cotton, everybody had a little money. Everybody was happy then. Didn't have all the conveniences, but everybody seemed to be happy. You never hear that nobody complained about one thing or the other. They didn't have it, but they adjusted to it. Peoples came to church better. They seemed happy, you know, they didn't have all the things but they could get to town. The Greyhound bus terminal would stop out there on the road right here, drop you off in Coolidge and drop you off in Randolph. You'd go in to church, catch the Greyhound bus and ride to Coolidge to church, and then come back home. They'd run often enough. That's when the church was in Coolidge.

Then when the church moved to Randolph, they didn't have to catch the bus to go to town to church. The church was right here. And the town was full of people then, men, women, and all. And in the fall, the people from all out of Texas, Oklahoma and Kansas, Missouri, and everywhere, they would just come in and just be church people. They'd pick cotton. All the stores flourished. Even Coolidge flourished. They had a lot of stores, a lot of things. But when all that went out, be like the town died. The people didn't spend the money, they weren't here.

So the machines for pickin' cotton hurt to a certain extent, it hurt the community. It didn't hurt the people so bad, but it hurt the community. Because all the merchants had a flourishin' business. And you didn't have to go to Casa Grande or somewhere way off to buy a lot of things. They had the stores, the department stores. You got to dress up in Coolidge. Three or four department stores. Three or four big grocery stores. A lot of service stations. A bar, a tavern, a little of everything. But machines come in and the people didn't come in from Texas and Oklahoma and them places anymore; well, there's nobody there to go to buy nothin' in Coolidge. So one store after another closed up. The hardware store closed down. The people toughed it out as long as they can, but no money come in and nobody had nothin' to go shoppin' with. People started leaving.

So I seen how it was with the cotton, when before the machines come in, the people picked cotton. They had money and a lot of activity throughout, from September throughout March. The machines, they'd get out there and pick all the cotton. They start pickin' in the middle or last of September, and before December they finished. If anybody make it anymore with money it's the farmer or financier. The people, some of them work on the ranch and make pretty good money, but they didn't make much on workin' the ranch, didn't make much money, but a lot of it was in circulation. You spent it where you lived. Back in the community.

When I run the store, when folks were on the field, I was busy over there. I was selling gas, pumping gas, selling meat and food. Inside I had a good little stock of groceries. Even bein' the pastor of the church. I didn't carry no canned goods, I carried some fresh meat. And everybody fared pretty good, you know. Didn't have a lot of money, but enough of them would get together and buy things. It didn't take that awful much.

My grandchildren now, I'm tryin' to get 'em to go to school. Tryin' to tell 'em, "Don't make too many mistakes, go to work, try for that house over there. If you can get it, take care of it," you know. "Don't be too loose, but don't be too stingy." The main thing is to help one another. You can survive if you help one another, because me and my wife could be in a lot better shape than we are. If my children need somethin', they havin' it rough, then I'm not satisfied. If your children are havin' a hard time and they hardly got any, well how can I eat, knowin' them grandchildren ain't got nothin'.

Not only do I just help my children and my family, anybody else that come to me need a hand I help 'em. So then sometime, I don't know, I got a lot of money out that I will never get back. But don't think about it. The thing about it is, what always aggravated me is, is when I get broke, there ain't nobody got nothin'! [Laughter] When I ain't got nothin', ain't nobody got nothin! But somehow or another, I stopped and settled down myself, then the thought come to me. You see sores. And every time that thought come to me I found what I want. I get what I need. But I don't crack down on people. I just give 'em my last, takin' my chance on some of 'em. But I always survive. So that way you live and you let live.

Give and it shall be given to you. Good measures, pressed down, shaken together. Good measures. Runnin' over. Lots more than what you give you'll be given back to. So this is a true sayin'. It's what Jesus did. He did, gave it to the poor. He lent it to God. Sometime I would go down the street. And when I see a man on the street holdin' "Will Work for Food," I looked at the man, and I pass him by. Somethin' says, "Go back." I stopped my car and backed it up—it was a colored guy. I backed up on North Seventh Street, and I give him a couple of dollars, I give it to him. "Thank you. Thank you." 'Cause I see a lot of people, some of 'em not tryin' to work. And then there's some in need. So you can't look at them and tell which is which.

I don't know what the future of Randolph is. It looks bad, but I guess it could be worse, you know. People got to have some pretty steady job, you know. Older folks don't want a whole lot of changes. And so you kinda adjust yourself to it. And sometimes you don't know how things are gonna work out, you know, but you stay with it, and somehow you know you'll survive. Not that you've done big things, but that you're surviving. And with the help of God you survive. He will fix a way for you to come out, you know. I didn't think I'd ever say that until I got in the church, you know.

You see in Coolidge vacant spots like here in Randolph. The buildings been taken down. Everything. Even the spark, even the joy in the people's eyes. They'd see the desolation in the city. One or two little bars is trying to strive up there. The younger people, some of 'em went out and got those jobs. One of my granddaughters got a job. But it's not makin' nothin' too much. It costs so much out there, it's hard for them to stay. It's tough for a lot of 'em. Some of 'em just working to pay rent.

And so, that's the way life is!

Metris White

Metris White comes from Wellington, Texas, and has lived in the Casa Grande Valley area since 1936, spending most of those years in Randolph. She came with her father and other siblings after her mother's poor health led to her death. I had not planned to interview Mrs. White originally, only her husband, Nick White, who had been a community activist, pastor, and businessperson there for about forty years.

But, as we talked on a hot July afternoon, Mrs. White became part of the conversation—adding, clarifying, laughing. She was very animated and funny and said some interesting things that were relevant to the story of this community. It was obvious that this seemingly quiet, unassuming woman had a great memory and sense of "the story." I then asked to tape her separately and did so on another visit. At that time, as you will see and hear in her narrative, he also graciously "helped" her. His and her narratives reinforce each other, reflecting each person's perspective and style of speaking.

These two wonderful people complement each other well and showed, in every way, their mutual respect. Neither tried to outdo, outtalk, or "out" the other. What Metris's narrative demonstrates is a meshing of voices. Also, during both tapings, several children, grandchildren, neighbors, and other relatives entered the room, yet both just waved a hand and kept on talking. Some of the grandchildren sat to listen to the stories.

On other occasions, as I visited them over the years in Randolph, Metris would tell me who had left town, who had returned, or who had died. She also saves obituaries and memorial service programs and shared these with me. They became a good source of information regarding place and family linkages.

When she exclaims in her narrative, "I need to write a history of what I know, I think I will! I think I will," she is absolutely conscious of her storytelling abilities. She has a wonderful ear for the effect of her narrative, as suggested by her intermittent laughter. If she does write her history, it would be a generous contribution to her family, community, and African Americans everywhere who can hear in her remembrances some of their life experiences; especially her early years being reared by her father, the agony of hunger, being without proper clothing, and living from hand to mouth under slavelike conditions in Texas and other places. The phrase "we grew up poor but didn't know it" is not appropriate for Metris's early years. She was poor, and she knew it. Her ability to laugh as she retells these experiences is a triumph of the spirit. At times her eyes would fill with tears and she would rock in place quietly.

I come to Randolph from Wellington, Texas, at the age of six years. Came in an ol' T Model. We had to take the top off of this ol' T Model before we could get over the hills. We came out in 1936. We had to push the car over the hills, it was in such bad shape. The first town we got to was Florence after we come out of the hills. There was six of us livin' in this one little adobe house with one room. No indoor bathrooms either. There was four children and my dad. There was another man come with him named Samuel. We came in November, and so cotton was gonna soon be over, and then we moved up here about three miles west of Randolph. Things was really tough; most of the time you didn't have a real good meal, you'd have a little, but not a real good meal. Just somethin' to say you ate somethin'.

When we come from Texas, my mother had just passed not too long ago.

Her health was bad, so my daddy figured if he could get us out here to finish raising, we'd have a chance at a better and a more healthy life. So this is one reason he come on to Arizona. My mother passed away in 1933. It was two of us girls, in the beginning, and then the baby girl come on. She lived in Dallas with my aunt, my mother's sister. So at, I would say 1939, she came to Arizona. Then she lived with us and then we moved away to Eloy. Stayed in Eloy, I guess, for about two or three years. We went back and forth from Randolph to Eloy.

To leave Texas and get here we had to pick cotton. We would stop and pick cotton all along the way to try to make enough money to come on here. That's how we got here. It was tough too. And some places we would go, the white man would ask us was we gonna stay. My daddy tell him, yeah. Maybe the first day he needs to draw out his money to buy groceries and gas, so he'd work then and draw it out and he'd leave. And we had to push this T Model Ford; we'd push it past that white man's house so he wouldn't catch us leavin'. And then when we'd get past so far, we'd jump in and take off. We did that more than one time, 'cause they wasn't gonna let us leave. The last place that we left, he stayed up all night seemed like it, but we got away, anyway.

Sometimes back then they'd want you to pick cotton until it was all gone, you know. Like maybe they might not get anybody else no way soon, so my daddy, he would play that trick when he need some money. He did need the money, you know, to buy gas and food. And so we just keep comin' 'cause I don't know now who told him about Randolph. Anyway, he said he figured if he got us here, since it was warm, we'd have a better chance because my mother was in poor health. And all in her family was in poor health. I was just a little girl.

During that time, they'd walk down the field behind ya on a horse. They never whipped you, but he was on a horse to see that you don't leave no cotton in that field. They was real strict back in that time. I tell you what, lots of people was treated that way back there in Texas. It was many, many people that worked and didn't get nothin'. They would gather all the cotton, just them ranchers wouldn't pay them nothin'. Buy them some groceries every now and then, but they didn't pay them nothin'. Of course, he didn't want to get caught up in that kinda thing. We worked hard when we was little. We could really grab that cotton. We didn't have to pick it out of the boll, just strip the stalk. I don't know how we learned it so easy, but we did. We could really pick cotton. So after I got grown, I could see why he would push the car past the highway. [Laughter]

And I tell you, we had to take the top off the car to come through the mountain. I guess it took us about a week to get here or better. 'Cause way in the night you would hear the animals over here. We'd go through Globe. *(Nick White: On Route 60.)* I don't know what it was. But it was a hard ride. We stopped in all them places. Livin' in all them. The mountains are so steep that one car would have to kinda pull over and let the other one squeeze by.

My daddy, he would get out there, and he and this other man, they'd make a big fire. It was a man at a service station told 'em to make a big fire so the animals wouldn't bother you. But one night they seen something. They thought it was a lion or a bear or something come real close to the car. And we got up and left, because we had no kinda shield to keep the animals from not attacking. And my oldest brother, he was very scared, he would always tell lies so he could jump in the car. [Laughter]

When my daddy got to Florence, he had thirty-eight cents in his pocket. So we barely made it to Florence. But, see then, you could buy a loaf of bread for ten cents. The other man didn't have no more than my daddy had, so one bought the loaf of bread and the other one bought some bologna. And that's what we had, sandwiches. I don't remember most things that happen last month, but I remember those days. It was very tough. When you go through somethin' like that, you don't forget it.

Not one soul did we know here. My daddy was a friendly man, so he soon met quite a few white peoples, you know. I didn't go to school. My first day in Borree Corner school I was ten years old. Most of the times, if you catch one schoolbus, then you miss the other. You get to where you supposed to catch the other one, it's gone! It was very hard.

We had all black teachers, you know, because it wasn't mixed. But it was mixed when it come time for me to go to school. They go to high school together, but they didn't get to go to grade school together. So, then we moved away, in '42 or '43, to Casa Grande. And we moved back to Randolph again. I'd say '43 or '44. It was more better schools in Casa Grande because here we just had one teacher; in Casa Grande there was two. We didn't have more than one teacher at a time, but almost every year it was a different teacher.

I didn't graduate from grade school, but I went to night school. I finally got a G.E.D. But in some areas, I didn't pass. I got disgusted and quit. But when I was grown and married I went to try to get my G.E.D. And that's all my education.

I picked cotton, sure did, and I could really pick it too! I guess I was between four and five years old when I started picking cotton. And I picked from then until even after I was married. And at the end of the week, you got paid for pickin' cotton on Saturday. You got it all at one time. The [mechanical] cotton pickers started pickin' all the cotton I guess in the '60s or '70s. I did some other work too. I lived in Phoenix, then we moved away to California, to Indio, but we wind up comin' back. We were lookin' for a farmin' area. They had pink corn, pink grapes, and I don't know what all. We did that maybe about eight or nine months. And then back to Randolph.

Pickin' that cotton was tough. But we had learned how to really grab it. We would make enough money to take us through the winter. Because there was nothin' here to do until the spring of the year, you know, when we start chopping. Sometime we work with contracts or the men that owned the cotton. My daddy, he never did his own contracting. And then later on, I don't know what year, he was kinda like blessed with gettin' his own farm, about three miles out, across the railroad. He did pretty good, and then somethin' happened to his well and he had to work on his well, and it didn't act right after that. We had to haul our water from way over in that area, where Mr. Fields and them live. *(Nick White: Yeah, we had to haul it in barrels. The barrels hold fifty gallons and a hundred gallons. Sometimes we'd have to pay to have water hauled, and it was a dollar a barrel to haul it here. See, there weren't but two wells here where the people used to draw water, and later on they dried up too.)* Later on it went to two dollars a barrel. And sometime that two barrels of water had to last you a week, you know, if you didn't have a way to get it. We'd have to keep anything from gettin' in except the mosquitos; you couldn't keep the mosquitos out of it.

At one time there was three grocery stores here. Sometimes things that we would need, we'd go over here and pick them up. But we'd go to Coolidge, Casa Grande, to do our shopping. But at one time there was four grocery stores here, and three service stations up on the highway, and then just across the street there we had a grocery store. First it was his [Nick's] daddy. Three of them owned it together. I had a brother owned one back down on the back street. And he ran an icehouse. But not when the Smiths had theirs. But the building is not there anymore. We sold the place here, and then we build over there, until one night Nick got tired of it. He got robbed. It was fairly good business. We had business from white and black, 'cause we would stay open a little later than the others. We had gas too, and that would bring business in. We sold bread, cakes, crackers, cheese, ice cream, canned goods, soaps, a lot of cheap things like that. We didn't sell any fruit. We had a meat vault and sold fresh meat.

One reason we quit the business was because the gas man would cheat us, you know. We'd pay off the gas bill, and he'd bring another one, which was supposed to be from the big drums or whatever you call it, under the ground. He cheated us, that's what he did. He wind up with the tanks, the drum under the ground that the gas in. We wind up givin' them to clear out what's there. See, he was aimin' for the land, but the land belonged to his [Nick's] dad, see, and his daddy didn't sign no papers for it. So the only thing he could do is just get the drum and go ahead with it.

Well, we bought the land from his [Nick's] uncle, his uncle that used to live here. His name was Simon Wandsey, the guy that sold the land that the Training Center is on, that land over there. *(Nick White: He had the first black store.)* Yeah, he had the first black store here, right there, across the street. He sold it about the same year, next year or two after we got married, so I would say it was in the 1940s. And I worked there at the Training Center for eighteen and a half years. I retired in 1984.

I met my husband in Coolidge. Oh, I don't even know what year that was. I think it was 1944. We got married in the forties, 1947.

I had four children. The first one was born here in this house. But I moved in this house here in 1949 or '50. I had the first child 'bout two or three months after we moved here, and I had a midwife, his [Nick's] aunt. And after that I had the other three in the hospital, two in Phoenix, one in Casa Grande. It was so much cheaper. I had a sister there, and I could go live with her where she was.

They were all raised here and went to school in Coolidge. At one time there was a school here. We had school in our church, but it was a different church. We got the same foundation, but the school burned down. Bunch of little boys burned the school down. We had no insurance. So my children didn't get a chance to pick cotton. [Laughter] Cotton pickin' machines got it. We stayed on here because, by that time, there was other work. My husband was doing construction work. So he would go up to, sometimes he go as far as Yuma and stay up there maybe two weekends and then he would come back. And now I just got the three here, one son and two daughters. The other son is in Pomona, California.

As far as blacks and whites was concerned, well, at one time it was kinda mixed up. I mean the blacks didn't want to live over on that side of the road anyway. They'd rather have been over here. So, way back then they owned places to go and drink and dance and do this sort of thing. They had three or four little joints that they went to here so they didn't have to mix with them. [Laughter] They had places that they would go to, and one place was on the highway. A little bar. A little dance place. And then on the highway over there by the service station. The Greyhound bus would stop here. It was a place to catch the bus and a place to get on. Now, the Greyhound bus don't come to Coolidge. So now, you have to go to Casa Grande to catch the Greyhound.

Some of the people would raise hogs here. We did too. We would sell hogs. Quite a few people raised hogs to sell. And we got some chickens. But everybody was happy. One didn't have a bunch more than the other.

We got along good with the white people here, too. Some of them, who had been for some time, was friends of my family. Those people that owned those grocery stores over where I was goin', well, they would give my daddy food to give us. We didn't have any money to buy any food. We would walk to Coolidge, and my daddy would pick us up.

I don't remember what kind of prison was behind the Training Center. It used to be a big prison there. Part of it was in Florence. When these prisoners found out what shape we was in, they would throw little things over the fence. And we wasn't scared. We didn't have sense enough to be scared. They would pitch us watermelon, cantaloupe, and such. They would wash our clothes. We would put them in a bag. Now you know we could have helped them folks escape and they wouldn't have asked us nothin'. My daddy always had a gun, and if they'd 've told us what to get, we'd a got it and whoop it over that top. Some things we didn't eat, you know. But they'd cook it and pitch it over to us. We lived in a tent, and we didn't have a cookstove. We'd cook it outside between two bricks or old rocks or somethin' like that. Maybe I need to write a history of what I know. I think I will! I think I will!

The prisoners, most of 'em just walk around out there. There really wasn't that many. Maybe something like about between I'd say twenty to fifty people. Minimum security is what it was. And see, I know the people in charge had to see 'em. We were only about six and seven then, but I remember that better than I think some things happen last month. When you really experience it, though, it's hard to forget it.

My daddy passed away in 1979. Yep, '79. 'Cause he was eighty-two years old. He had lost the farm around the end of the sixties. But the man that did it, he didn't have no good luck after that. The guy that dug the well, they didn't want to see him do good. His cotton was comin' out good. He had owned it a long time. Homesteaded. And some of them same guys helped him to get that land and told him about it.

I think our children have stayed here because we live here too. That tall boy, he moved to Yuma, and they just been back here, I guess about six years. He wanted to work at the church. Then my daughter, she don't belong either, but they want to come here and work for the church. There's quite a few young children around here, but it's not like it used to be. But I tell you one thing, since Miss Mathis up there passed away, almost everybody that moved away from here come back. That's one thing, they'll come back to the funerals, you know. Some of them live in Sacramento, California, and way up there like that. They come back to the funeral. Yeah, they do. Most of the time everybody that once lived here come back.

Well, I could tell you some really funny things. My daddy would keep us two pair of coveralls apiece. Made in one piece, you know. And when it be time to go to bed, well he would wash them, you know, and get up the next day and we put on the other pair. So it got to the place we didn't have two pair, so one of these white women, their name was Cockerill, that was their last name, took in chicken-feed sack bags, it was really pretty, orange and yellow, so they sewed my sister a yellow dress out of those seed sacks.

And you wouldn't see, at that time, white people lettin' little dirty black kids eat at their table, but they know what shape we was in. They wouldn't let us in, not when they eat, but after they eat, then we could go on in and eat. Now we didn't know nothin' about no prejudice or nothin'. We were just eatin'. [Laughter]

My children like for me to tell them stories about that. When they was young, I didn't mention it too much. The big son, he really laughed about it. You know, if he'd have been there, he wouldn't have laughed.

Ron and Pearl Jordan

Pearl Jordan (McKinney) is the eldest child from the first marriage of cotton contractor Cleveland "Preacher" Jordan of Hope, Arkansas, and Nona Jordan. Both came to Arizona in 1941. A subsequent divorce or separation ended their marriage. Ron is the eldest child from the second marriage of Cleveland Jordan, to Myrtle Williams, a Pima Indian woman from Sacaton on the Gila River Reservation. The children of both families have maintained a close relationship, which I understand happened because the second Mrs. Jordan raised most of the first set of Jordan children along with those from her marriage to Cleveland.

The dynamics of the conversation with them were very interesting as each corrected the other from time to time or wanted to tell the other's story. Pearl spoke sparingly about her personal life as migrant, daughter, sister, and woman. They both seemed genuinely happy and caught up in their memories of the old days—the cotton days of their early life and early Randolph. Their relationship displayed a special connection as well as respect. I had met Pearl at another time, but I thought the dual conversation was like a call and response and exhibited interesting dynamics as the legacy of two family histories converged.

P: When we left Arkansas, there wasn't too much work for my father, Cleveland Jordan, so he just came by here and made his home. We went two months without work. There was some little somethin' goin' on all the time, from irrigatin' and pickin' cotton and chopping cotton. I think we came about '42. Maybe summer of '41? That's a long time. That's fifty years. This is 1992. The whole family came, my mother and father and my grandpa and his children. There was only four or five houses in Randolph. There was nothin' out there but just desert. So we was just livin' in little tents. And he was pickin' cotton, you know, farmin' out there near Casa Grande. He decided to leave that and come to Randolph and build a little ol' camp. He'd go to the Arkansas area and bring hams back to feed the workers. He done a good job and he worked us all. He did a good job, I think. We were satisfied with like a quarter; and we thought we had it made! He kept us workin' even when we grown up. Well, the good thing is you could come home on the weekends and get to look at your boyfriends. [Laughter] That's the good part of it. And the bad part of it was when you had to go back out there to go to work. Parents were strict in that time, you know, they didn't "court" too much at that time.

R: Actually, what I can recollect, you know, is Pearl and them comin' up as a child, most of the courtin' was done in the field. When you worked you didn't have a whole lot of social time. They went from sunup to sundown, so if you met somebody it was usually one of the hands that they brought from Arkansas or somebody locally and you socialized in the field. Your main socializing, if you went, was at church. They did give you time to go to church. So if you wanted to meet somebody that's gonna read the Good Book and be decent, you went to church. With Pearl and all the sisters in the family, daddy was very possessive. He didn't want his girls goin' anyplace. Most all the girls ended up just takin' off and met somebody. Like they say, got that chicken away from the hen. [Laughter] And that's how it was kinda done. I don't recall any of my sisters ever getting married or having a wedding or anything like that, formally. Everybody just met somebody and went to the Justice of the Peace and got married. We listened to the fights when Joe Louis and all them was fightin'. We'd all sit around the radio and listen. Even the people that worked for us. So it was little time socializing. We was either workin' in the fields or workin' in the house. Our father was very strict. He kept us jumpin'. He kept you busy. The yard wouldn't look bad. He'd have you out cleanin' up. He prized his yard and his flowers and trees and stuff. He kept everything neat, and you was part of it.

P: I know when he worked the hands, he'd give them all a day to get groceries. He loaded up all his hands and go to the store and get groceries. Took him about a half a day doin' that. Then he'd bring 'em back home.

Well, let's see, back in them times our older brother, we call him "Woostie," but his real name was Grover, married another girl that was from another farm and they all worked together right there in Randolph. The Taylors was a big family. So that was a marriage that happened, and they just continued on workin' just like we always done. And Willie married his girlfriend, and then her family worked doin' farmwork. But, after Willie, things started changin'. Automation came and started takin' the jobs. That was the only kind of work you could do around here—labor type work. We couldn't go to town and work at the gas station or work at a fast food or hamburger place. You just didn't get hired. It just wasn't thought of. The only thing you could do was go out here and shovel dirt so the water could run down them roads or work on the farm or take care of somebody's cattle or irrigate. That's when the water came through; that's about in 1960 somethin'. So everybody was lookin' for a better livelihood. So a lot of us left. We got the water down that road about one-quarter mile. On the right-hand side is a cotton gin. And they would let us get water there. So we would haul our water from that gin, which would be about two and a half miles. People come with fifty-five gallon drums, and get water there, and that's how we got our water supply.

R: Now my father had the best water tank that there was beyond that camp. In Randolph they had to pay to haul them barrels of water. That wouldn't be done but twice a week. My grandpa was a water man there. And he would go, and they would pay about two or three dollars a barrel to bring it. Yeah, my daddy didn't charge. Just my grandpa here charged 'em and the others, you know. We had a tank. We had a big tank. It was mounted on an old trailer. And we filled that tank up. Thinkin' back now, that tank would hold probably four hundred gallons of water. It was a pretty good-sized tank. And that tank provided water for everybody. We had a little faucet built on the back of it so you could just come and get your own water out of it. So when it would run low we would just hook it up to the truck and go down there. That was part of my job and Willie's. I was small, but in that big truck, that's where I learned how to drive. Willie'd be sleeping . . . we'd do this 'til two or three in the morning. And he would teach me how to drive. And that was fun. We didn't have running water, so to take a bath we had a number two or three tub and you'd heat water on the stove and mix a little cool to get the temperature right, and then you'd jump in that tub, and when you'd get in you'd throw the water out. We'd wash our faces, you know, we had washpans back in them times, and we'd wash our face. We all might wash in the same water. [Laughter] We didn't care, we'd get a towel and wipe up and that's how you do it.

P: If we work, we work a whole week. We started on at three dollars a day. Then it went up to five dollars a day. And so, that five dollars a day goin' like twenty-five or thirty dollars a week. So we had to buy our food out of that, you know. We had to manage with that. That was good days then. We never did have no big problems, just knowin' when that money comes. Then we go shopping on the weekends on Saturdays. And then Sunday we'd just have a good meal.

R: To kind of answer your question; how did they keep track of the money and everything? See like, that was my dad's business. He contracted out, so naturally he had ledgers and books and then he would just write down names. For example, if it was Mr. Jones over here. He had so many children workin'. You keep track of who was who. They'd write your name down and what days you worked, and whatever, so you'd get paid accordingly. If you didn't come to work, you didn't get paid. It was not like now, sick leave or annual leave stuff like that. When they plant cotton, they planted so many seeds you had to thin that cotton out. When you got paid, you was rewarded by the acre. So when you go out there, you go up and down those rows, you had to chop, thin out, maybe ten, it just depends on the length of the rows. About ten rows or twelve rows to an acre. . . . So some people, the younger people, the stronger people, even the ladies, was good at that.

P: Naturally, the more you did the more money you made. But then we did it another way too, where it was good for everybody. Some of the elders still needed money, and so we'd just work at a certain pace and get paid by the day. You put in your ten hours and you got paid by what you did.

R: That was when the cotton was first comin' up. Now, as it got on up into different stages, we did different things. As it got maybe through the growing process, weeds and different things would grow in it. Then we worked primarily by the day then because there was no rewards for tryin' to go through and work an acre because there may not be too many weeds.

P: If that one acre has weeds you'd just work that one acre all day. Some farmers went along with this acre and some of them just go on and pay 'em for the cotton and they put in an acre. They would just go by the acre.

R: Dad would negotiate with the farmer, and they would come to an agreement like I mentioned. Well, the last I remember it was seven dollars a day for ten hours of work. Say I come to you, you are a contractor. I need a job. I'm a good hand, but the most the contractor would pay me is six dollars a day because he made the other dollar for his profit for the expense of the trucks and water and other things like that. So back in those days it was very beneficial for a family to have eight, nine, or ten kids because when they work those eight, nine, ten kids if they was workin' age, or able to work, they got six dollars per head. So the family made good money. Now, Pearl said we didn't have rough times. But we did. We worked during the harvesting season, but when that harvesting season was over, daddy would buy flour, lard, margarine, rice, and all the essentials that we would need to get through the winter. 'Cause there was hardly any work durin' the winter. We'd stock up, because there was no welfare. There was no food stamps. There was no nothin'.

P: The store would let you charge sometime to the workers.

R: Or the farmer loaned you money until work started up again, and you were in his debt. But then, there was good clean water in that lake. And uh, durin' the wintertime my dad and a lot of parents, they went fishin'. And they caught beautiful, good-eatin' fish. And so, as greedy kids, durin' the hard times, we ate up all the meat. We had salt pork. If they didn't catch any fish, we didn't have fish to go with rice or beef that evenin'. We were sent to school with just rice and fish many times. We thought we was simply poor, but we was eatin' good!

P: Daddy'd get on that truck, and he would bring our food at school. We would sit on the back of the truck and eat our lunch. And then he'd go back home. And when he go back home, he'd have a job for us for eatin' that fish too. [Laughter] Sometime he'd bring lunch in a lard bucket. Little metal cans. They knocked holes in the top for ventilation, and that was your lunch pail, and put some biscuits and salt pork. Sometimes we'd get slices of an apple or orange.

R: Sometimes my daddy, he'd go behind, even the farmer. He'd come out to check and see what's goin' on. Sometime when you hire a new person and the new person come in, he might say he knows how to cut the weeds properly or do this or that properly, but you don't know until you get him out there and turn him loose. If you don't know, Dad would show you what he wanted done. And if you were cooperative, you would go on through. Most everybody was cooperative, 'cause they needed to work. But sometimes you'd get some rascal come through that didn't want to do right, and after a while you'd have to let him go. But it was trial and error over a long period of time. People were pretty good people. Some maybe didn't have the ability to do it, but they did pretty good. If Daddy couldn't be there to do the work, he might drive over here to talk with another farmer to get more work for us. So he would appoint somebody else to be the second boss, or what we call 'em, a "straw boss." He'd be the straw boss. Usually, when I came up in the fields I did a lot of the extra runnin', goin' to the bank, did errands and stuff like that. He would always want to make me the straw boss. Here I was young and drivin' that bus and that truck and tellin' grown people what to do. "Come on, line up this way!" and "take this line" and tellin' 'em, "Well, take your time, it's gettin' close to eatin'. Slow down." "Stack them hoes, let's go." And all they'd do is just put the hoe in the soil and leave it standin' up. And we'd all go back to the bus or truck and find a place on the ground or underneath the cotton.

But, you know, the heat didn't bother you then, because it would be 105 degrees, you'd wear a hat, you'd be sweaty out there and them breezes come up out there and with your sweat, it would cool you off! Sometimes if you got too far from a tree, we'd load the workers up and take them to these trees. Cottonwood trees or wherever. We'd all sit under it. We'd get up under the truck or cotton trailer, sit there and eat. When he said, "Let's go, it's time to go back at one o'clock," we go.

We worked ten-hour days, but it wasn't like now; you work two hours and take a ten-minute break and get a half hour, hour lunch . . . there was none of that. You had time to go drink water or use the bathroom. That was it. Well, sometimes I was a waterboy; me and my other brother Bennie. We carried buckets of water. You had to take the bugs and leaves out. [Laughter] But that's how it was done.

P: Well, some people now got a good job workin' at the Trainin' Center. And some of us didn't get a chance to get a job out there, you know, so they had to take off and leave. And there's still a length of time when you have to find a way of gettin' out here and find 'em a job. Education didn't have nothin' too much to do with it when they was hiring at the Arizona Training Center. Some of them didn't have an education, and they got to stay on. Now if they had to quit, they wouldn't be able to get back on again. They were just basically lookin' for workers, the laborer type. So they can just put people to take care of the handicapped ones. But then once it got up to where there was requirements, you do have to have certain qualifications. I have thought about goin' back to Arkansas to live, and I have gone back there. But I couldn't stay there 'cause I got used to bein' here.

IV. “We Are Sojourner’s Daughters”

The Casa Grande Valley Cotton Women

"Life for women cotton pickers who had to be a wife, a mother, and a worker was very hard. Some women would be picking cotton until they near 'bout had a baby in the field." —Parlee Bradley

Passing by the Randolph Community Baptist Church on King Street, by the Church of God in Christ Pentecostal Church on Bell Street, and by the local grocery, which boasts a poster of a beautiful black woman touting COBRA MALT LIQUOR, I feel a deep kinship with this place. Visible from any street in Randolph are yards filled with cars from ten, twenty, thirty, forty years ago, abandoned homes, trailers, the debris from years of work and life. In the yards, clotheslines are filled with children's clothes, working people's clothes, clothes that tell more than the wearers of them ever imagined, tired-looking clothes, drab-colored clothes, cold clothes, nonflapping-in-the-wind clothes, old defeated clothes. Someone told me that when she passed by Randolph on the way to the Hohokam ruins near Coolidge, she saw quilts on a clothesline. Yet, despite my many visits, I never saw any quilts on a line or in a house, although I know they are there. This is not to say that the women whose strong hands picked cotton could not have lovingly created beautiful quilts; after all, historically, black women's quilts were some of their most unique and precious possessions. A son told me that his mother, indeed, made quilts, but that he had them at his home in Casa Grande. The lives of Randolph's women themselves, however, form a symphony of colored quilts—unique, memorable, and nonduplicative.

The women of Randolph, Arizona, are a special breed, and I have always been intrigued by their "secret lives." Their histories must be rich, fascinating, revealing, hard, passionate. I know it is the women who continue whatever traditions there are and carry the tales, the facts, the records of this village.

Randolph is a village, and the women are its griots, keeping the village community, the big yard, the compound, alive. They know who everybody is, where everybody is, who is related to whom; they know each bloodline, kinship, quarrel, and concern. They know whose child walks the road, where each child lives, whether they have eaten or not, and whether they ought to "git on back home to yo' mama." They mother this village like their African sisters historically have done. There is a lot of relatedness here, linkages, family trees, kinships that never change whether one stays or leaves the "village," whether one returns to it in disgrace or in glory. Randolph, home, is the mothering place.

The women of this village came from Oklahoma, Arkansas, or Texas when their men or parents, as part of the later exodus in the 1940s and 1950s, came to harvest cotton. Like the braceros, the entire family picked, hands ruined young, going behind the machines that slowly took away those jobs; picking "blossoms," pooling their money for food and essentials; saving money to buy pads for when their "ladies' time" came upon them, squatting in the fields, lacking feminine space and privacy.

They went to church meetings, hoped, prayed for a better situation, jobs, and education for their children. Their young girls grew up maturing like the cotton, "befo' time," full of desire, some marrying a Randolph man and starting families. Then, cotton didn't need them anymore. The hands, parched from the sun, became idle. Clocks stopped. Everything stopped. The crippling effects of all of these things brought the village together. With no place to go, no skills to market, they stayed, and, like the Okies, they huddled together. They became one body, with one heart and one head. Some of the women sought work on white farms; some went further to do domestic work, often not seeing their children or families regularly. Reminiscent of the black woman's place in the community of slaves, the old usually cared for the "young un's."

The late 1950s and early 1960s were critical for Randolph's women because the first migration away from Randolph was of the men to nearby cities—Phoenix, Tucson—then further west to Los Angeles, Seattle, and Portland. The second migration was of young women to similar cities, perhaps looking, subconsciously, for a prince. But they returned more often and more quickly than their men, realizing that with no skills or education, a city is a hostile, unwelcoming beast.

Some were smart, talented, beautiful, ambitious, and wanted to go to college. But parents declined. Nice girls did not leave home. Or, the parents were afraid to lose them to the beast.

So here they sit, aging, looking older than their seventeen, thirty, forty-five, seventy-one, seventy-nine years—women left alone to care for children and grandchildren, read the Bible, go to the American Legion and play Bingo. Some work at the Arizona Training Center, but in most cases these are menial and dead-end jobs. The good jobs in surrounding communities—Eloy, Coolidge, Florence—hardly ever fall into their laps. "You are overqualified," "you are underqualified," "no blacks wanted today." What to do, where to go?

Randolph's women have no chapters in Arizona history, no chapter in the history of the Civil Rights movement, they have no paragraphs or sentences in the developing history of the women's movement. They have no streets named for their heroes. No Harriet Tubman Street, no Sojourner Truth, no Mary McLeod Bethune. No Angela Davis street, no Rosa Parks, no Assata Shakur, no Fannie Lou Hamer, no Coretta Scott King. Somehow, the historic events associated with those women never touched them directly. Yet they carry in their entire being the spirit of those sisters. The years rush by on the old highway like the fancy cars. Like a black town in the desert, nothing seems to fit or make sense anymore.

But there's a wonderful feeling of bonding and nurturing between these women. There is a strong silken thread that holds all their lives together. Not everybody loves everybody, but they are caretakers of each other's lives. A grandmother walks wearing "out of place and time" clothes, a wool hat with a summer dress, trailing seven kids behind her like Harriet Tubman leading her troops to freedom. She demands their obedience and, like Sojourner Truth, does it all. Her life has been hard, but her gait is sure. She is a crucible of strength, continuity, and common sense, a true black matriarch passing on that strength, discipline, belonging, and pretend childhood joy to them. But her real thoughts are: "Lord, let this child have more power and money than I have had so that my name shall be written in riches and power, so that my child's strength shall be blackness in my gray hair and firmness in my feeble steps and a monument to my name."

Their "sometimes working" mothers toil elsewhere as the white man's "slave," suffering indignities, laughing behind their masters' backs, thinking "we shall overcome" thoughts, bringing home another day's imaginings.

In the old days, up until 1995, at the local store tended by the woman overseer, Mandy Moore, they would take off their woolen caps, buy "Dark and Lovely" perms, hot sauce, and treats for the kids with an "I'll pay you tomorrow, got to get this child's head done" smile. Besides church, the store was the meeting place. Like the clothes on the lines, what you buy and how often you buy it tells all your "business" in a town that has no secrets. At the store, they used to put together the quilting pieces of the day or week or month. The men dash in and, with guarded greetings, buy cigarettes and beer. But the women convene to discuss the day's agenda and the day's worries, and to fantasize soap operas on Carmichael, Malcolm, King, and Kennedy Streets, speaking in their Mahalia Jackson, Sarah Vaughan, Bessie Smith, and Aretha Franklin voices.

> Blackwater blues done tole me to pack my things and go
> Cause my house fell down, can't live there no more.
> Umm-mm, but I can't move no more.
> Their ain't no place for a poor girl to go.

These women don't cry. The tears are in their voices, in the direct vacant gaze, the walk, the strong shoulders muscled from work, the glorious voices at church on Sunday, where hope always gets reborn.

> I'm gonna lay down my burden
> Down by the river side
> Down by the river side
> Gonna lay my burden down. . . .

There are hundreds of stories waiting to be told, hundreds of stories waiting to be written, hundreds of feelings waiting to be unleashed. Randolph is the western Brewster Place. Living in this "walled-in" compound, rocking with their "deferred dreams" intact, they wait for change, dreaming dreams that shrivel "like raisins in the sun."

Strong, spiritual, determined, patient women; long livin' women; "beloved women"; "ain't I a woman" women; "color purple" women; "everyday use" women. Women with "nothin' to do and no place to go," cause Randolph is where her folks "is."[1]

One historian states:

> In recent years the black woman has almost become a Romantic, legendary figure in this society, because the vast conceptions of her as a person are largely dictated by these stereotypes. The idea that she is almost superhuman, capable of assuming all major responsibilities for sustaining herself and her family through economic and social conditions, has been projected in much of the popular literature as well as academic research. . . . Of Black womanhood on all fronts . . . she is felt to be stronger than other women and certainly stronger than Black men.[2]

This statement is true to some extent, in this town, but these women are strong partners to their menfolk, young and old, and raise and praise their children of both genders as well as support their sisters. The women's community here is strong, and some "worked so hard and suffered so," without major complaint. During slavery, "Black women were not considered 'weak' females, but more like beasts of burden. They were not included in the 'delicate female' image or certainly the 'American housewife' image. Sojourner Truth, formerly a slave and activist in the abolition and women's rights movements, eloquently expressed the differences in treatment of Black and white women under slavery."[3] Sojourner said: "That man over there says that women need to be helped into carriages and lifted over ditches, and to have the best place everywhere. Nobody ever helped me into carriages or over puddles, or gives me any best place . . . and ain't I a woman?"[4] Sojourner's spirit and creed live on in her "daughters in the dust."

no ♀ needs that kind of paternalistic help

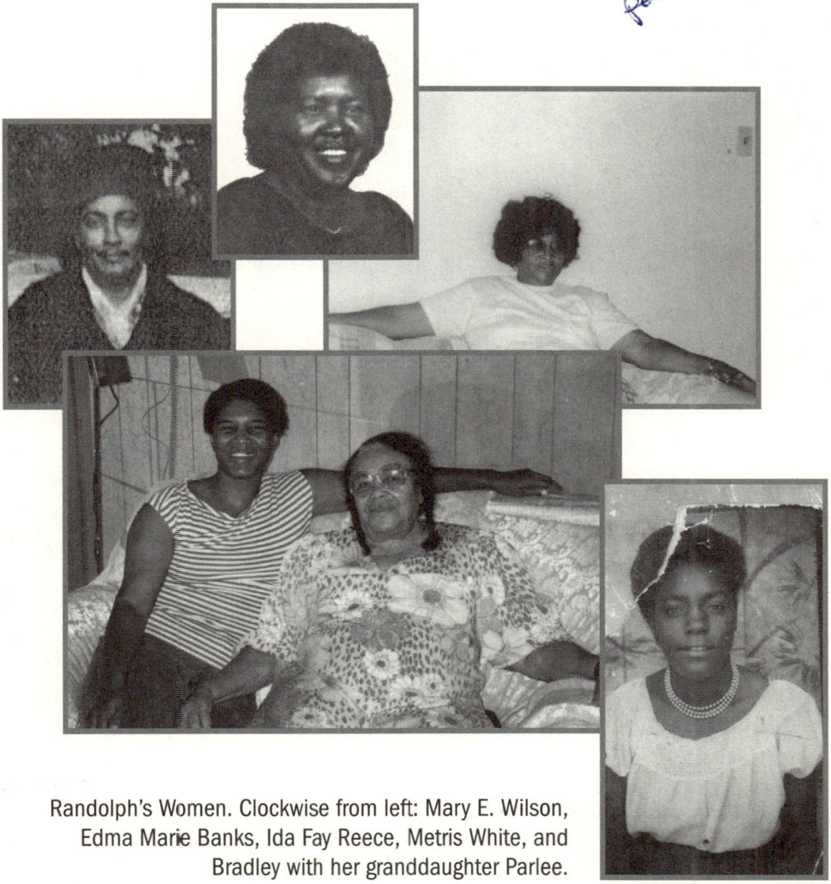

Randolph's Women. Clockwise from left: Mary E. Wilson, Edma Marie Banks, Ida Fay Reece, Metris White, and Bradley with her granddaughter Parlee.

Parlee Bradley

I met Parlee Bradley after reading the article "Cotton, King of Fibers" and seeing her picture in the June 1994 issue of National Geographic. *This was a special issue on the history of cotton, universally. The authors of the article found Parlee after someone at the Eloy City Hall recommended her and her husband, James, as representative of the Negro migrant workers in the Casa Grande Valley. The Bradleys had done cotton work for at least thirty years and had remained in the community. She says, "They wanted me to go and show them the place where we farmed, but there wasn't farming no more in them places. They took pictures of us at the little house we lived in when we first came out here." She adds, "I know more about cotton than most people."*

Mrs. Bradley is very proud of being an international celebrity because of the National Geographic *essay and pleased that I was interested enough to locate and interview her. Parlee, born in 1914 in Arkadelphia, Arkansas, was eighty-four in 1998 but does not look her age. She speaks proudly of being the granddaughter of a Malaga Spanish grandmother and a Congo Negro. She remembers stories about slavery told to her and her siblings by her grandmother. She is a handsome, light-skinned woman with long dark braids, a strong body, a delightful laugh, a determined face, a businesslike manner, and a voice that resonates, yet is animated. She is a most engaging person to be around. On most visits, her granddaughter, also named Parlee, would be present. Their closeness shows, and the younger Parlee seems quietly protective of her grandmother.*

This exuberant woman returned to school at the age of fifty-five, after work in the fields diminished, and received an associate of arts (A.A.) degree from Central Arizona Community College. She believes it's never too late to better yourself and seeks to set an example for the young ones. She has always been a fighter for people's rights and challenges anyone or any institution that seeks to hinder access for and equality to the sons and daughters of slaves. Parlee has been "fighting" folks and the system a long time and tells how she and her sister beat up a white girl who took their pig from their yard, back in Arkadelphia. She also talks about the KKK warnings her family received because they refused to back down to whites on many occasions. She took on Pinal County, the town of Eloy, the state, and the federal government for educational assistance and work opportunities. She is active in many community groups and causes, including the NAACP, Expanded Food and Nutrition programs, Head Start, and the Migrant Opportunity Program Board. She sees the Migrant Opportunity Program as of critical importance in the valley for those like her and her husband, James, "who had been pushed out of the fields." She fully uses her associate degree in elementary education, child development, and accounting by doing income taxes for people in the community. She is as sharp as a whip and shows no signs of slowing down despite what she says about quitting all those activities. She feels that if she lets a cause go, no one else will carry it on, and so she refuses to let the "white folks" take over.

She is the only narrator who spoke at length about the issues of poor migrant black women, their place in cotton picking work, the problems of inadequate prenatal and postnatal care, child-rearing difficulties, and displacement by the machine. She states: "Life for women cotton pickers who had to be a wife, a mother, and a worker was very hard. . . . [S]ome of them would have their babies on the end of the field. . . . They didn't even fix a decent place for them to lay." Parlee Bradley has always been a visionary, a progressive woman much ahead of the third wave of woman warriors agitating for change.

Her narrative is laced with colorful stories of Eloy—black, Mexican, and white.

She also told me about Georgia Rupe, that fascinating early black successful businesswoman who came to Eloy with nothing but a sack of clothes and determination in her eyes. Eloy needs its own book, telling its history from a minority or migrant perspective. For this volume, I have edited Parlee's story to keep the focus closer to her personal experiences, thoughts, and opinions; and she has had many.

My name is Parlee Bradley, and my family, my husband, and children came to the state of Arizona in 1946. I'm from Arkansas; my husband's home was in Texas. We stayed together fifty-four years. We married in Texas. We lived in Arkansas for a while, and then we came to Texas and sharecropped. We'd make crops. At that particular time people was paying up around fifty cents a day for labor.

We would grow cotton, corn, peas, pumpkins, and other vegetables. All we had mostly is what we'd grow in the fields. And we'd raise chickens and hogs and things like that. At that time if we would clear a hundred dollars at the end of the year, we were lucky. That was in 1932. But before then, during the Depression, we didn't make that much.

We were in Arkadelphia, Arkansas, close around Little Rock where they grow cotton and different things. When we came to Texas, we heard about Arizona. They said you could make money, money, money. At least that's what they told us.

Some friends we knew was coming out here in a semi-truck and hauling migrant workers who would work in the fields in the state of Arizona. He didn't have too many people, so we put all our furniture and everything in the semi-truck, and my family, my husband's sister, and her family all came out here. And when we got here it was at night, and they put us off in a little old place where there was no running water.

They had some little cabins sitting there with no windows and a door to go in and a little shed you'd raise up where you could see the light in the daytime. And there I was with all of my furniture I'd brought from Texas sitting out under a tree. We couldn't even get it in the little cabin, and so they put a tent up we could use. We didn't have coolers, no ice trays, and by the way that was the time the war was ending. And we had stamps to buy food with. We couldn't even buy meat in Eloy, Arizona. It was in the summertime and it was real hot and the people we was working for had cattle and they grew cotton too. So some way or another they run a calf into the fence and he broke his neck and we had some meat. We had money now, but you couldn't buy no meat. There was rationing. You had stamps to buy sugar, you had stamps to buy meat, and so forth. I remember coming to Eloy; they had one grocery store, and we did find some skins; I guess that came from Mexico or somewhere. One time we enjoyed those pork skins. We went and fried that out for the grease. And we could eat the skin. We had money, but we didn't have no meat. We didn't have no refrigerators, no coolers, nothing. You could buy ice at a little place, but we had to take a canvas sack and put the ice in there and put it in the ground and cover it with dirt, and that's how we'd keep our meat.

We was working for two brothers; their names were Ray Thomas and Lester Thomas. They're the one had leased this farm and was farming and raising cattle. These are white men. I would chop cotton, and my husband would drive the tractors and irrigate. Then my daughter and son, we'd be chopping cotton and we'd be so hot. We'd be walking down the road, and I'd heard stories telling that the ants would pick up sprolls and put under them to crawl across the sand. I got to see it was true. They do that. There's a lizard would do that too. You know put sticks, with their legs and hold them, and wobble across on them. We'd be walking down where the pump was and water was running; they'd have an outhouse there and we'd be walking along and you could just hear rattlesnakes, just rattling all in the water. And then finally one day we found one in the little cabin we lived in. Oh, honey, it was somethin' else. We stayed there and worked like that for years.

My one child was nine and the other was seven. My sister-in-law, she had a little girl. We all stayed in those cabins. And then in Eloy, we would come to town, but people mostly lived out on camps. There weren't too many people lived around. What you see now wasn't nothing but desert. But they'd all come to that little town. And, oh, it would be so terrible. They'd be shooting people; people on the streets with guns. I guess you heard about how open Eloy was. Somebody would get killed every weekend. Whenever those people come off the camp, they'd come and they'd be a shooting every weekend. That's the way it was, it was very violent. People coming from everywhere, all kind of people—white men, Indians. It would be blacks shooting each other, whites shooting. Well, I saw a black man kill a white man. I saw a white man kill a black man. It'd just be killing each other. Didn't matter about the race. They'd be drinking, they would be gambling on the streets and drinking and there'd be bars. There'd be prostitution. Just name it. You know, they wrote a book on Eloy. On the weekend, that's when they'd be kicking in all these bars, prostituting. Shooting and fighting and killing. Oh, it was terrible. I wish you could get that Eloy book. It tells how they found women embalmed in the cupboards. And then other real bad things. A black man killed a judge. Violent stuff. It's a whole lot. Oh, it's terrible. You know you ought to just get it, and that'll tell the whole history of the people. Yeah, it's called the Eloy book.

Then we traveled around to different fields and ride in different trucks and work for different contractors. And later on down through the years, we bought us a car, and I think we was about the first one had a car. And we'd had to take so many people to the hospital every weekend. They would come to us and if somebody died in Florence, they would call us because we carried so many people that was hurt over there. It was just an outlaw town. On the streets you could just see everything you wanted on the streets. Without going in a place. The streets of Eloy was pretty colorful. You'd see people with guns, with thirty-eights on each side. I think some of these bad types were escaped criminals seeking shelter in migrant work. The landowners never asked us questions on backgrounds except, "How much can you pick?" When the guy waited for the sheriff, they killed him right down here on Seventh Street in a little café. And then the judge, the white guy killed the judge that sentenced him to the jail down there on the streets in Eloy. So much happened to people. And even people come from England to see this place.

They got a bar down here; the Mexicans run it. There's about two or three around here. They have a killing down there a time or two. Soon it quit being wild.

My husband wasn't too rowdy. But he come out here and he'd go to bars and stuff. But I'd go and peep and go on to the picture show and come back home. He mostly worked all the time, until he got disabled. When all of this was going on, I was going to school and all. And when he retired they didn't give him but ninety-eight dollars a month. He retired from migrant labor. Most of the time when you worked they didn't take out social security. But they kept on raising it each time, but we couldn't make it on that. The ninety-eight dollars he'd get wouldn't pay all the utilities. He was sick, and I had to go out and go to work, and then the Migrant Opportunity Program come along, and that's what give me a job. We paid our home off and could take care of things. But I made it.

Then the government took over and made them go to building better places for people to live. After going through those years my children went to school. They had to go to a segregated school, so we all joined the NAACP and marched on the schools till we got it integrated with Mr. Bud Cooksey. He was the NAACP president. Then when we got through, they integrated the schools here. We worked for that. We's all members, but he started it. After that life got a little better. We had a cooler, and so we worked one fall, and I bought a lot to build a house on. Me and my husband and children; we put it up brick by brick. We'd build one room at a time till we got us a house where we could live in. We stopped living on the camps where all the killing and stuff went on. And from then on, we kinda prospered and done very well. We contracted cotton sometimes. My children growed up, and they left and went to California. Then me and my husband was still here, so I went to working in the vegetables and things. Wasn't no cotton to pick, wasn't no cotton to chop, for they made choppers that chopped the cotton. So it just wasn't no work to do in fields 'cause the machines took over. The machines came along about 1959. And by the '60s, we didn't have no cotton to pick, no cotton to chop.

Well, sometime you could get a job weeding a little cotton that the choppers couldn't get. And then they got machines burning the weeds with chemicals. Now they still weed cotton a little bit. We had to chop it, thin it out one to two stalks every lick. Later on they learned how to grow cotton where you didn't have to do nothing but weed it. And then they eventually learned how to plant it where you didn't have to thin it at all.

Then the government comes in with a program for migrant workers, so I went down to get a job there. There was a schoolteacher here, Troy Thomas. He started the program. He was very instrumental in getting people that worked on the farms and didn't have no jobs with the Migrant Opportunity Program. If you was a migrant worker and you could work in day-care centers, they would send you to school. So I went to work for them. The next thing I did was to be a recruiter for education, to send young people to different states to get their high school diploma or to go to college, and furthermore, I was going to school too. I went to Central Arizona College and took courses from the University of Arizona until I got my A.A. degree. First of all, then, I went as a community coordinator going out seeing to the needs of the people, and it was bad. Old people were trying to go to the field and wasn't nothing for them to do. And they wouldn't put them on welfare. So my job was to go around and see what was needed, then take them and see that they got whatever help they needed. This started in 1966 here in Eloy, and it is still going on.

The Migrant Opportunity Program gave us the center here. I worked with them, then after that I decided to get a building for black families. I was on the Migrant Opportunity Board in Phoenix, and I got enough contacts with the Council of Churches to build a day-care center here in Eloy. I was on the board when we went and built this building. And we was going to get fired, all of us. They wasn't supposed to *build* the building, they were supposed to *rent* a building. And so when we did that, we were in trouble. I went to the Council of Churches in Mesa, and they went to the Great Southwest, a financial company, and they paid them off $60,000 for this building. Then we rented the building for $690 a month for them to use for a day-care center, until we paid the $60,000 off. And too, I started a second-hand store and paid for about

What was this? research paper topic?

Great Society

six lots. And we sold second-hand clothes, and the Catholic Diocese in Tucson furnished clothes for us to sell until we paid for the land. That's how we got this place out here. It's just the blacks and we own it. And so we got a nonprofit organization. I got that with the commission in Phoenix. Then we pull out from the Migrant Opportunity and become Community Action. That was in Randolph, Eloy, Coolidge, and Florence. So when we went to Community Action, then we went on our own. We pulled out from the federal. Well, it was federal funded, but it was a different program. And then, when we all retired, Head Start rented it. That's when there was a colored school in Head Start. After we leased the building to Head Start, I got a job teaching there. That's where I retired from.

Children couldn't go to the high school here in Eloy because it was all white. During this time, working for the government, I integrated the city—the whole city and the housing. See, you couldn't cross the track here in Eloy. You had to live in Joe Louis housing, you couldn't go across the track. I even tried to rent a house across there and couldn't. And so we started working on that, and my kids got really scared for me for a while. They thought something was going to happen.

So we did get a building for the community down there. We did get that. And it did, you know, I didn't have nothing hardly until I went to work for the Migrant Opportunity Program. After that I went to Community Action Commission. I was an accountant. I took up accounting. And so I been doing taxes after I retired. And just this past year I did income tax. I don't think I'm going to do it next year, though, I'm going to quit. And so it really did help me, that program really did help. And that's what I said about young people today. If they try, they can get an education. Look, I was fifty-five when I went to school. I went to school and got all kinds of different things that I could do. And now I can make extra money. If you could get it free and why not take it! The government was furnishing it for the migrant workers. But it's mostly Spanish people now. You know there's no blacks working in no fields. The Spanish, they come in there and pick lettuce and pecans and grapes, and oranges around Phoenix.

The Indians picked and chopped, too. We worked with them. And when we would go and pick peaches and pick vegetables—we'd be working with the Indians. But they didn't stay with us long, they liked to be to theirselves. You know they don't like to be with other people. They lived in camps by theirselves. The Indians were a quiet people, and they didn't ever be in too much trouble. No more than getting drunk. The Indian men didn't flirt with the black women. But the black men flirted with the Indian women. That's what I couldn't understand. But I often watched. In living here all these years, I hardly ever saw an Indian with a black woman. I only know of one black woman that an Indian went with. And he came from Coolidge to Eloy to see this lady. And that's the only one I ever knowed. But, ooh, the Mexicans, the Spanish people, they flirt with everybody. The Indians they come here, and I still talk with them. They come in, I fill out their papers. I help.

When we first come out here they would pay about a dollar a hundred for picking cotton. Then it went to three dollars a hundred. I believe that was as high as I ever got. I picked around six hundred a day. Couldn't nobody beat me picking cotton! There was about four or five people could do it. I was one. I could really pick that cotton! I think that's what broke my husband down and what got me crippled now. Doing all that hard work. And we would chop by the day. Sometime they got to six dollars a day. And then they would pay by the row. And if you got paid by the row, I could chop fast so I could make good money. Around twelve dollars a day because I could really work fast. But if they paid you by the day you couldn't hardly make anything. As long as I've been here it would be good times and bad times and sometime you would be making no money. But all through those years, it looked like I was so happy. I didn't have anything, but I didn't have nothing to worry. We'd all be in the fields singing songs and talking about the Bible and just happy and didn't go to bed 'til after twelve o'clock because it was too hot.

In the off season we'd go to California, Utah, or Colorado. We'd go and make it back in time to chop cotton. We'd put down a hoe from chopping and pick up a sack and go to picking. But when we first came out here you didn't have to go nowhere because they growed stub cotton. Stub cotton is where they just cut the stalks and let it come up from the roots that was already growing. And for years when they growed a lot of cotton here, we didn't have to go nowhere. We had a year-round job picking and chopping cotton. Because when we quit chopping, the stub cotton was ready to pick. So we'd pick that, and when that was gone, the other cotton was ready to pick.

They grew pima and all that kind of cotton. But like it'd be different season for it. The stub cotton would come off in the time this other cotton would be growing, maturing. And so that kept us working in the fields all the time; for years it was like that. They quit growing stub cotton, and then that's when we had to go to different places to work until the cotton picking come open. We quit chopping cotton before we quit picking cotton, because they quit thinning it. They learned how to let it grow thick so the cotton pickers could pick it better.

People who pick cotton got all kinds of ailments; bending over all those years. I think their back and their knees mostly be hurtin'. We not only had to bend and pick cotton off the stalk, but when the machines come in they didn't have the root machines, and we had to crawl on our knees and pick it up off the ground. Well, the machines picked it and it would fall out a lot on the ground. And then they would hire people to go in. That's all the kind of work we'd get to do in the cotton picking field is go pick that up off the ground. After the machines went over it, we'd go back, we'd put knee pads on and be down there picking it up. Next thing they got root machines and they went and picked it up. You know it's ridiculous, they always got something to keep you from doing the work. Yes, for a good many years we picked it off the ground. So it's hard on your knees and your back. Well, it was hard on a lot of people's hands, but I wore gloves all the time. You know I picked with gloves. But it didn't seem to bother my hands or my back then. Later on I could feel that that was the cause of some of my problems. But I was awful healthy and worked real hard.

We went a time or two to Utah and Colorado when we had a car of our own. And the other times they'd be hauling people. They'd come in just like these migrant people come into here to pick cotton. Trucks would come in here from Utah and Montana and California and pick up people that wasn't working at that particular time and take them to gather the crops there and bring them back when the time came. But then we lived in little places when we'd go up there. They'd have army barracks for everybody to live in. They didn't have the little camps like they had when we first came to Arizona. Some, they stayed to irrigate.

When cotton kind of went out and machines came in women still needed to work. But I didn't like housework. I never did go work in people's homes or nothing. But some did. I went to working with the vegetables. At Santa Cruz we picked mustard greens, onions, carrots, we did all that kind of work. We'd bunch carrots in bundles, and they'd pay us so much a crate. At Santa Cruz Farm, they would grow all kind of vegetables. We also picked pecans. They grew potatoes, too, but not too many in Arizona. This would be in the fall of the year. That's when I was picking those greens. You had to wear rubber boots, and it'd be wet all the time. You had to have slickers on. And we'd be so cold out there early before daylight! It didn't matter how early you'd start, because you got paid by the crate anyway. There wouldn't be too many blacks out there working either. Just a few. And we'd work in the lettuce. Some would be on the machine mostly. It used to be they could do it by hand. Then there'd be two or three on a machine. And then they'd box them. That was the only other kind of work the women would be doing. And they had a shed here where they'd haul the lettuce and greens and things, call it the packinghouse. That's gone now. Nothing much on a farm they could do. But instead of having a shed now they truck them out. I guess they got one in Casa Grande somewhere, maybe Coolidge, that haul them out to pack. Now women didn't really do too much domestic work. Lately they went to building all these truck stops and cafeterias and freeways and then now the women mostly go work at those places. They had a sewing factory here too, but it went out.

All the business is out on the freeway now. The freeway missed Eloy. All them places is out there—hotels and motels. And a lot of them go to Casa Grande to work. And the Training Center, that's where most of the Coolidge people work. It's over there by Randolph.

My children picked too until they got kinda grown. They went to school and got out of those fields. They went and learned something different. My son learned how to be a cement man and a plasterer. Then he run his own business. He'd go in and stucco houses, mostly plastering. And then my daughter, she married, and her husband was a truck driver so they all left this place and went to California.

The Mexicans just got it all. There's a few Indians working on the farms now, but not too many. And since the Indians, they got so much money from the government they stay on their reservations. They got water and stuff so they're making a pretty good living. Well, now back out here in the little Choo-Choo reservation, they live pretty good. They farm. They have factories up there and training and a police force. They also live pretty good up here at Sacaton. I know of three reservations. There's one at Maricopa, and there's one out back here from Casa Grande. And there's one out from Coolidge. I think its called Blackwater.

Life for women cotton pickers who had to be a wife, a mother, and a worker was very hard. Some women would be picking cotton until they near 'bout had a baby in the field. They would. And some of them would have their babies on the end of the field. That was terrible. Some of them when they got where they could have cars, they'd have them all in the cars sitting on the end of the row. The farmers and men didn't pay you a bit of attention. They'd be out there ready to have their baby, be picking cotton one week, and have the baby the next week. But the contractors would let somebody empty their sacks. They wouldn't let them crawl up on the trailers on the ladders and empty sacks. They wouldn't say nothing like "take a couple of days off." But sometimes the contractor would help. He may take them home sometimes if they got sick. But the foremens were stern. They didn't even fix a decent place for them to lay. They didn't care.

Some of the black women were midwives. There were only two I know of. They had Mother Brown and I can't think of the other woman's name. Well anyway, I know they waited on mine. One was waiting on my daughter, and I had to get down and get that baby myself right in this room. Oh, she wasn't doing anything. And they didn't have no doctor here in Eloy. And no hospital in Casa Grande. I think there's always been some kind of hospital in Florence. They had a horse doctor here. He'd doctor on people. You know he was a veterinarian. I'd take my kids to him. And he'd help them too. That's when they first knowed about penicillin. When they had a flu or a cold, he'd get to them. But I mind he wasn't a real doctor. He'd go to doctor our horses and cows and people too. Dr. O'Brien, I never will forget him. I guess a lot of people dies on account of they couldn't get to medical help when they need it. After she have her baby, my daughter-in-law and daughter stayed about two weeks away from the fields. And then my daughter and my daughter-in-law, they would get people to keep their babies. They wouldn't take their babies to the field. You know there was a lady keep them. I know her name was Miss Willis. They couldn't pay her much, like twenty-five cents a day. This would be an older woman who couldn't go to the fields. So they would leave the babies there with her. And some of them would leave their kids with the older kids. And the kids wouldn't be that much older either, some of them would leave them with kids about eight years old.

I always planned on returning to Oklahoma. That's the reason I have the kind of house I'm living in now. I always planned to go back and build me a house. But I waited so long when I did go back there, I didn't want to go back there anymore. It was all growed up, and there was so much changes that I didn't want to go back there and live. But all the time I did dream of going back. There wasn't even farming where we came from anymore. They couldn't even grow cotton and stuff like we had back there. And all the old people, they died out. The farms are sold and people just turned it into cattle and stuff. So after about twenty-five or some years, we didn't want to go back. The only reason why I hung around was because my husband could drive any kind of machine so he went to driving cotton picking machinery. And that's why he had a job until he got disabled. Women and children that worked in the fields didn't have no jobs. Some of the men stayed because they could drive tractors and irrigate. A lot of them was gone. And eventually there just wasn't no black people working on the farms. Ain't none working on there now.

We have been in this house since 1952. It had to be before that, I guess it was in the 1940s. It has five rooms. I built two bedrooms, kitchen, dining room, bath. We just kept building on it. I didn't try to build a better house because I was intending to go back home. When I went back there in twenty years, you couldn't walk through the woods. It grew up just like a jungle. And so I didn't want no part of it, I was scared to even get out and walk. When we left there you could roam through the woods and fish. I didn't like it at all. When I went back, it had changed. So, I'll just stay here and see to things as long as I can. See to the kids and the womenfolk.

Georgia Rupe

For years I had seen, in various texts, a picture taken by photojournalist Dorothea Lange, who chronicled migrant life of the Depression years. The picture is of a billboard in Arizona with three signs on it. One is for the governor's election: "Jerrie W. Lee for Governor"; another reads: "Cotton Pickers Wanted—11 1/2 mi. So. mi. E.—Curly Fraley's 'Camp.'" But the one that had fascinated me the most was the one nailed on top of the other two, which reads, "Wanted, 75 Colored Cotton Pickers—75—Rupe's Cafe—1/4 mile north." I've always wondered: Who is Rupe? Was Rupe black or white? Male or female? Where is this café? Could it still be there? Whom can I ask? I did surmise that it was somewhere in the Eloy, Casa Grande, or Coolidge area, but where?

It was my luck, therefore, to meet Parlee Bradley, who knew about not only Rupe's Café but also Georgia Rupe, the proprietor, and Eloy's first black migrant woman and entrepreneur. Rupe, with her one-armed husband and three children, landed on Highway 84 with a cotton sack and no money and became the material of many stories about her business ventures, affiliations, and place in Eloy's colorful history. I had unlocked the mystery of that sign after all these years, but there was a lot more to learn about the Rupe family.

Courtesy National Archives

Eloy gained a reputation as a tough cowboy town, the site of many murders and gunfights and a haven for criminals. Deputy Sheriff Jim Sloter is the author of the only book written about Eloy's bloody history, entitled Eloy. *The book begins: "Eloy, the tombstone of today boasts of killings. It's known from coast to coast as the toughest town in the United States and the West's last frontier. . . . A town with such a name as ours, Eloy, which means 'my God.'" In the book, Sloter profiles three black men, Alonzo Davis Case, Fred Blair Case, and Willie Bishop—all hardworking cotton pickers whose addictions for or love of women and liquor led to their violent deaths. There is also a picture of the black section of town, Joe Louis Heights and Frontier Street, where Georgia Rupe had her café, as well as the Dandy Tavern. In the book, thirty-year-old men are called "colored boys," and "colored girls" are described as victims of circumstances—poverty, class, "absent" men, alcohol, violence, and gender oppression.*

This is the context and environment in which Georgia Rupe made her mark. She survived by crossing over class lines, race lines, and gender lines, from cotton picker to businesswoman and power player on Main Street. Parlee Bradley's remembering of her is reverential and poetic. Georgia Rupe remains a special woman and a heroine for those blacks who came to Eloy for field work and kept hope alive.

The following narrative about Georgia Rupe is adapted from Parlee Bradley's memories of this fascinating woman.

Rupe's Cafe and Rooming House, in Eloy.

Georgia Rupe and Charley Rupe came from Oklahoma, I believe Idabel, Oklahoma. They had three children. Two girls and one boy. They were almost grown. They came along about 1936. Georgia Rupe was the first black woman in Eloy! And when the truck put them off on Route 84, the one that come through Eloy just dropped them off with a cotton sack with their clothes in it. Then she walked on down and looked and didn't see nothing but desert. She walked about five miles to Friendly Corner and found a camp. And that's where her and her children got them a little cabin, and that's where they lived. But after she worked a while in the fields, she came and rented a place and opened up a café and went to selling food and adobe. Well, they picked cotton for about six months, and then they opened up this little place. First she went to selling sandwiches in the field, and then after she made enough money she seen it was good to open up a café, a tavern, whatever you want to call it. She started selling food, and people would come far and near eating her food. And she made good. Then when the people went to moving in from the South and different places, they wanted to separate blacks from the whites, so she had to sell. She was forced to sell her little tavern across town. They sold her some land over here. Mr. A. Treckle, he sold to her, and furthermore, he told her he'd just almost give her the whole Joe Louis Heights.

Joe Louis Heights starts from the railroad crossing—that was the line. The other side of the tracks was the City of Eloy, and back where the black people started building, they called it Joe Louis Heights. Mama Rupe started that because she was the first landowner here. Then after she went to making money, she built a hotel and made money from that, and from there she went to selling lots. She sold nearly everybody a home over here. They bought lots from her. They was selling from $100 to $150 a lot. People would put tents, little cabins, whatever they could on them. Down through the years, they went to building houses. Then she bought trucks and made her son a contractor; to get workers to do cotton chopping and picking cotton. So then he went to contracting, hauling people from other states back to the state of Arizona, transporting them back and forth from Texas, Oklahoma, and Arkansas and all those places. He'd bring them in the fall of the year to pick cotton. She built some cabins for them to live in around her hotel, and those that he bring in would move out on the farms. She sold her tavern across town and moved it over here to Joe Louis Heights. And so she built up a reputation in the community. That was her lifestyle. She went into real estate and got to be a wealthy lady. Nearly everything we have, even our churches and things, we bought that land from her. I bought this place from her. Most everybody that own land here, they bought it from her. From about 1940 on up.

All of her children is dead. Her daughter-in-law, she took over and she has some nieces and grandchildren here. She was a very businesslike woman and was a good person to my eyes. I liked her; she knew how to manage. She just come in here in a truck, with a sack with clothes, and was put out on Highway 84. She was a kinda heavy-set, chunky, brown-skinned lady. Looked like an Indian.

Her husband was one-armed, he had got hurt in the army, and he would just help her with the hotel. She was the businesswoman; he'd just be there all the time. He couldn't do much work, and was I think drawing a veteran's check or something like that when she come here. The whites didn't bother her too much. After she left from over there on their side of the tracks, they bought her out over there. By then they paid her a good lot of money to move out from over that way—the white side. And they didn't never bother her; they kinda respected her all down through the years. I used to go set down and talk to her, and she'd enjoy talking with me. At first she cared for the hotel. But later on she changed and joined the church and she helped us build Mount Zion Baptist Church.

I don't know if Georgia loaned money, but I know she took in money. I know one time that my children had lots and they left and they didn't pay for the lots and she gave me a chance to pay for it. She was kind of good about that. She'd give you time to pay for it. I don't think she'd let you get too much in her debt, but she wouldn't take the lots from you. She was good like that, at least for me she was. And I think she probably helped a lot of people, too. She seemed to be a kind woman.

Her daughter-in-law, Mary Wilson, went to selling lots. She took over all the property. She was running the hotel till she died here recently. They sell to renters now. Like a rooming house where people rent apartments. It used to be rooms by the night, and she always had it full; you could always go there and get a room in those days, but all that's gone now. It ain't the same no more. She was kinda older when she came and old when she died round about 1990.

Yes, Georgia Rupe came out here and started from scratch and built herself up. And when she died, she had lots of money. She left her family a legacy. She left them land, houses and stuff, even a farm in Casa Grande. Mrs. Rupe is an example of a person who went from "sack to riches."

Fannie Reece

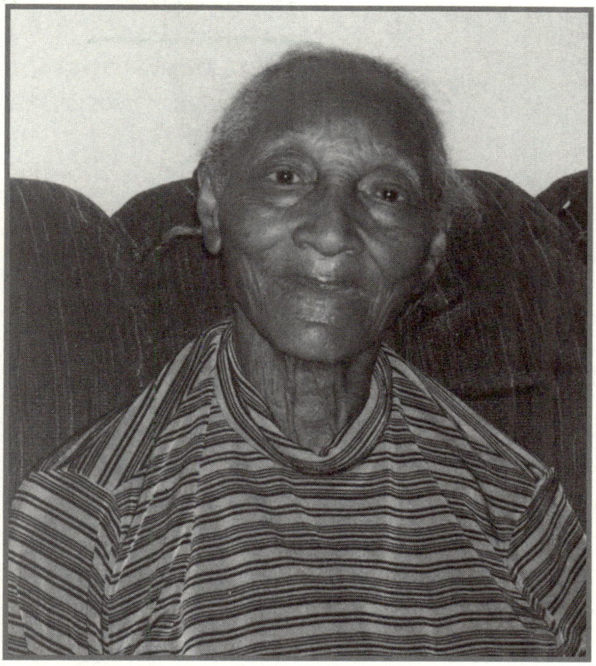

I think I have known Fannie Reece longer than any of the other residents in Randolph. I first saw a picture of her, holding her ten-month-old grandson, Victor, in the Arizona Daily Star *in 1980. She is also the Sojourner Truth of my essay written for* Compendium *magazine in February 1989. While I was visiting Mandy Moore in her grocery store, this tall, thin woman with a slow but determined walk came into the shop to purchase some items. Behind her were several children—her charges—perhaps seven of them. She immediately reminded me of my grandmother Arabella—the honey color, the erect shoulders,*

Courtesy Arizona Daily Star

the brown twinkling eyes, the firm touch, the control and authority she projected. We spoke little. She said, "Hello, ma'am." I responded likewise. As she left the store, my eyes followed her as she walked back to her house on King Street with all of those children following her. She, indeed, to me was Sojourner Truth directing her children to a loving sanctuary—her house. I immediately knew she was special. The major part of her narrative was done in 1991. We have simply talked and visited after that.

Ms. Fannie, I soon found out, is one of the most revered women in Randolph. Although she is not the oldest woman there, when this interview was done she was seen as the village matriarch. Our relationship grew to frequent visits just because I liked being in her presence, and she seemed genuine about missing me when I did not come. Over the years, I have met several of her children: Ida Fay; Zola Fay; Curtis; L. B., the once Golden Gloves champion; Aaron Jr.; and the eldest daughter, Jesse, the bright one, who looks a lot like her mother and wanted to be a schoolteacher. As Ms. Fannie's health declined, I visited her in hospitals and at home. Throughout all of these low times she kept her optimism about life. Her "faith in the Good Lord" is admirable. She is never alone in her home as various sons, daughters, and grandchildren take turns "sitting" with her—night and day.

Although homebound for the past few years, Ms. Fannie is still the center of activities in her King Street home. Her eighty-something voice still rings out when she says to a grandson, "I done tole you not to butt in when grownups is talkin'!" I have taken several pictures of her on every visit in order to record her journey through illnesses and birthdays, as well as the grandchildren's quick maturing and ripening—especially the once-young Victor, who today is still her protector and source of joy. They are so close that one needs only to glance from

Fannie Reece and her grandson Victor at the age of twelve.

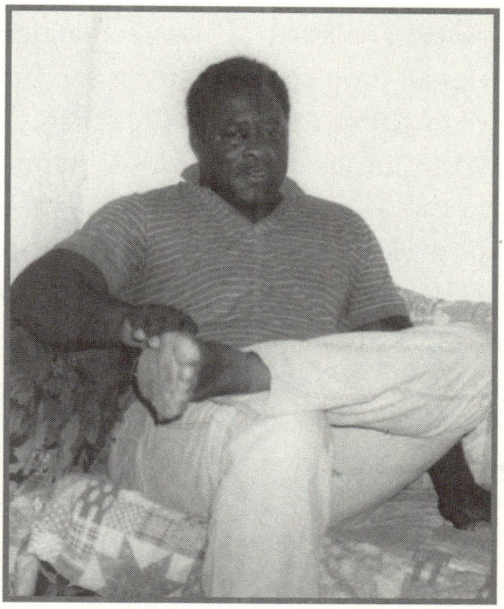

Aaron Reece Jr., son of Ms. Fanny.

one face to the other to recognize a silent language between them that needs no verbalization.

She always addressed me as "Ma'am"; she always seemed delighted when I came and wistful when I left, so I always promised to return.

I'm Fannie Reece, and I came to Arizona on the truck to pick cotton. My grandmother was half white and part Indian on my mother's side and my grandfather Negro and Cree Indian on my father's side. He was real black. My uncle looked like a white man. I came from Hugo, Oklahoma. Hugo is two hundred miles from Paris, Texas, and one hundred miles from Arkansas. But the community where we live at was in Fog Bottom, the part they call Little Dixie. Down there we mostly grow cotton, corn, peanuts, chicken, cows, and hogs. I was raised on a farm. I liked it back there.

I left my childrens with my mother back in Hugo and was supposed to go back to get them, but I got so big [pregnant] I couldn't go. I bring the baby Aaron with me though, and soon the others done come. I had thirteen children, and nine are alive now. I lost two in Oklahoma and two in Randolph. That's right, nine. Two live in California, one in Casa Grande, four in Coolidge, and two here in Randolph. I was born September 10, 1917. Gettin' old now. My husband's name is Aaron Reece. He came from Pittsburg, Texas. He died in 1974.

I had two sisters out here in Randolph. They didn't come to pick cotton but to live. I have now two sisters and one brother. The onliest two black people here when we come was Mrs. Moore and Mr. White, Nick White's father. When we was pickin' cotton we worked under Melvin's mother. I remember young Melvin [Moore] running around behind her on the cotton trolley. We worked under Mr. Jordan and Sister Clara Bell too. They was contractors. We lived in Clara's camp houses. When we came here seem like it was mostly Indian, and then they died out after the black people come to pick and chop cotton. Randolph was bigger when we come here. Some left for California to retire, you know. Some run off to find work when cotton gone. Me and my husband came in 1950, and the first place we stay at was in the mountains. A place called Chandler Heights. We stayed in a tent on the camp of a white contractor. This place here we bought from Mrs. Youtsey for about $350.00 round about 1952 or '53. By 1954, the same lots was selling for $600.00. My first house burned down in 1971. My sister Fuddie was a smoker, and she started the fire. We lost just about everything—pictures, marriage license, a mink coat that was gave to me, and my false teeth. It took a while to get more teeth. The house got remodeled in this same spot. There's a bunch of Mexicans in here now. Started to comin' and buyin' about five or six years ago. Some are farmhand squatters. One next to me right here, came in about a month now. But I be nice to 'em.

I only got to the third grade because I had to babysit while my mother worked. Those were the days when no welfare was around. We had to work for what we got. My mother was a sweet lady. A sweet, sweet lady. I used to do piece quilts and really enjoyed that. I used to cook a lot too. My husband didn't like cornbread. He liked biscuits. So I cooked two types of bread. I'm too crippled up now to do any a that anymore. I used to have a garden. Grew okra and greens. But after I had my surgery in 1975 I couldn't do it no more. Hurt to bend over, you know.

I belong to the Church of God in Christ on the corner of Bell Street. This is King Street. Nick White is the pastor there. The young people and children don't even go to Sunday school no more. I don' know where they goin' with their lives these days. My grandson Victor Ellis, I been raisin' him for a long time. I hope he do right. His mother, Zola Fay, lives in Coolidge. I worked hard raisin' kids. Had seven welfare children one time.

(The following was said by Aaron Jr. about his mother to extend her story.)

> My mom, she worked hard in the field. And she even worked sick in the field. She practically took care of us. Sometimes we wondered where we gonna get the next meal, but some way she'd always come up with a meal when we get home. I remember when she was sick one day, she worked. Then she was laid out a little bit and then she get up. She tried to make a day. So finally one day she just got where she couldn't work no more. So me and my brother L. B., we picked up the slack there. We picked cotton. Sometime I picked 300 and 400 pounds a day. Sometime L. B. would pick 500 or 600. We would give her all that money. We would keep what we make on Saturdays. I guess after that she was put on some kind of welfare. She was drawin' some kind of a check off of Johnnie L. So we made it. My dad, he just sat at home, and when time got hard he would get out in the field and irrigate, or get out and drive tractors.
>
> But Mother got to where she couldn't work no more, so she just got down. The girls, they had gone out on their own. So we had like Johnnie L., Curtis, and Zola Fay. Most of the reason I quit school, you know, was 'cause she couldn't work no more. But we had to kinda take care of ourself. And sometimes we help pay our mother always. We pick cotton, so she would always put that money in the bank. She would save for us. We'd come up with some food, we didn't know where it come from. I guess she would save the money we would give her durin' the time we were pickin' cotton. We would get all the money on Saturday. So I guess she saved that money and when time really got bad she would buy groceries with that. So my three sisters right here, they all picked cotton and raised us. That's the way we survived, you know. It was pretty hard then.
>
> That's how we got through this. Other than that, it would have been rougher. Then we all grown up, got bigger, most of us . . . time got better, we could get a job, so we made it for ourself and help her. My mother would always find a way to feed us.

The Lord has blessed me to take care of all of them. Took care of my husband, took care of my sister and brother-in-law. Saw my two kids die too. I say as long as you can do what you can do for one, then you feel free.

Mamie Selmon

Mamie Selmon, the mother of shopkeeper Mandy Selmon Moore, was perhaps the oldest living woman in Randolph when I met her. My conversations with her were held in 1991, when she was eighty-one years old. We only "visited" after that, but her conversations were always filled with the militant voice of one who had seen a lifetime of wrongs done to her people. She related incidents from her childhood, at the age of perhaps eight to ten, when she was growing up in Oklahoma. She knew of the stealings, lynchings, beatings, and senseless and unjust acts done to Negroes and Indians by whites. Mamie is a mixture of both bloods. These experiences left her wary of the dominant culture, and that mistrust has lingered into her mature years.

She was not one of the hundreds of cotton women who labored year after year in the fields of the Casa Grande Valley, but, interestingly, in her living room are decorative items like wreaths and flowers made from raw cotton puffs. Perhaps

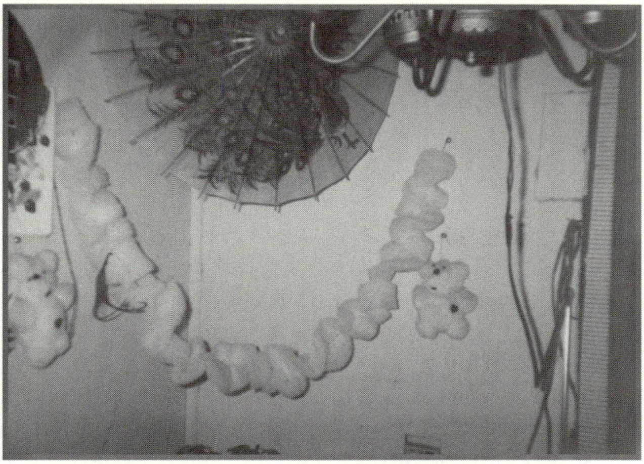

Decorative cotton items in the living room of Mamie Selmon.

even though she grew up around cotton and respected those who worked with it, that's as close as she wished to be connected to it. Her perspective on Randolph is realistic as she comments on life in the town and on world events, the future of the young, and what was and what could be done to better social and racial relationships. She is surrounded always by her daughters and more than sixty-seven grandchildren and great-grandchildren. She is loved by all of them, but she doesn't stand for any nonsense from anyone, kin or not.

My name is Mamie Selmon, and I was born February 5, 1910. I came to Randolph in 1954 from Eufaula, Oklahoma, with my four kids. I'm one-half Indian-Choctaw. I came here for my health. My husband was Virgil Selmon, and he was born in Texas. He came to pick cotton in Oklahoma around Clinton and Eufaula. That's how we met. After cotton got eaten up we decided to leave.

My sons Jacob and Herbert really came to Arizona as adventurers, not to pick cotton. One Saturday night, in Eufaula, they just showed up in the yard with an old covered truck. I said, "What you think you doin'?" He say, "Mama, we goin' to Arizona, there's work out there." We loaded up that truck and took off the next mornin'. I had nine children, total, from two husbands. The first set of six was Sells, and the second set of three is Selmon.

I was raised in Oklahoma among Indians and whites and saw a bunch a bad stuff there. The whites killed a black and blamed two blacks; they lynched him. I never forgot that. One time a white woman threatened to slap me. I said, "Slap me and they'll find you dead!" Here in Randolph when I was workin' at the Training Center a white man talked about hurtin' me; I say to him, "You do it and they'd be an ambulance here to pick you up." I don't like white folks, no kind a way. Take me a little time and I'm gone. I don't want nothin' to do with 'em.

I didn't like it when we first came out and there wasn't no white folks in Randolph. One moved in about four years ago. I don't know where they came from. My first house was across the street, and the lady who sold it to us didn't give a good deal. We bought it for $100 and sold it for $500. Melvin's mother had a lot of homes; I bought one from her about 1967. It had no floor, so my brother and son worked on it. We put in a cement floor and walls. Now the Mexicans is buying up black folks' homes. My sons started messin' around with local girls and petty soon got to marryin'! My husband worked as a handyman and did some building. He picked some cotton, but I never did pick cotton.

Nick White, the pastor of the Holiness Church, had the first store down the street from Miss Fannie, and Big Mama White, Nick's mother, had control of everything 'til it died out. The Griffins and the Woods along the highway had a store earlier. They white people. Even though I didn't pick cotton, I did hear stories about women goin' into labor in the fields. How could they carry a child to date like that!

I worked at the Training Center for six weeks but left because they wanted to treat black folks bad. I just wouldn't stand for that. Most white people leave their children and kinfolks there to die. One child been there for so long she now forty-five years old. I had a day care in Randolph from 1967 to 1976 when it went out of business. We lost the land and the building in what I call "a great land swindle." We is still gettin' cheated outa things we worked hard for; yeah, the swindles is still goin' on.

Beulah Moore and Christine Moore Fields

Christine Moore Fields.

I found Beulah Moore's legacy so intriguing that I felt I had to attempt to reconstruct a part of her life, as it is important to the Randolph story. She came to Randolph in 1944 and died at age sixty-two. I had met her two sons and one daughter. Her presence in terms of landownership and the cotton contracting business was a first in Arizona history. She was one of only a handful of black labor contractors, of any gender, in the Casa Grande Valley community and the only woman contractor in Randolph. She shared with her "sister contractor," Georgia Rupe, the strength and determination to pursue part of the American dream and exhibited the black pioneering spirit of the West. Not only that, she did it without a mate. So hers is a truly heroic and unique story.

Her daughter, Christine Moore, who married a Fields (Samuel), spoke about her mother guardedly, perhaps thinking of the modest, unpretentious woman Beulah Moore was. Another interviewee said the following:

> *Melvin's mother, she contract. She contract at the time we chop it; contracted the pickin' too. She was a very hard-workin' woman. She saved her money too. She didn't blow it. She was very nice. She was always tryin' to help the people's lives. She would contract so people would have some jobs. Mostly women folks, you know. She'd drive a long way to chop it and pick it too. She had her own trucks, but just one truck at a time. They came from Arkansas.*

We came out here in 1944, and I was about eleven years old. Well, we was originally from Hope, Arkansas. And my mother came out here, and we really didn't know anybody. She had a cousin that was already out here. We came on the train. And before we got to Texas we saw many Indians perched on the hill. I saw Indians, and boy that scared me to death. That's where the train stopped, and all of those Indians were there. And in those days they wore long dresses and they had long hair and the silk blouses and I was scared to death. All them, that's the only thing I can remember about it. I was scared to death. In Arkansas, there weren't any Indians. I never heard of 'em before. They just were scary. I remember the coach was all black and we'd hide when the Indians got on. Some other people came out here. They got off in Tucson. I don't remember their names.

We stayed on a farm about two miles east of Coolidge, me, my brother, my mother and us. There was three boys and me. My brothers' names were Melvin, Billy, and Claude. Me and my oldest brother, he's dead now, he and I worked. We was too small to care for ourselves, so we'd both have to work and chop cotton and rode together. It was real hard back in those days. And hot. They didn't have coolers and things. Oh, it was rough. My mother decided that since my dad was in the service at the time she could do better out here. And that was after my grandmother died. So we came out here. My mother always been a hard worker. She chopped cotton and picked cotton, and during the off-season she'd do domestic work for white folks in Arkansas and here too.

We just always had to work. Whenever we wanted somethin', she'd make us work and save your money. If you wanted anything, you'd have to work for it and save up to buy it. And I went to school here in several different places. The schools then wasn't integrated like they are now. So we was way out in the country, an old country school. One room, two rooms. From there my mother wanted a place of our own, so we bought a place in Randolph. That's where my brother, Melvin, stays. The one his son (my nephew) lives in now is the original house. Well the post office used to be further down, goin' east. But they built a new one. And most of the white people did live over on the west side. I don't know why; but the house we bought was bought from white people. It was small, but, you know, it served its purpose. In those days you made it in one room or two rooms. and you didn't think about it.

And so, after a while I just didn't feel like goin' to work and goin' to school and my mother's out strugglin' and so. And my brothers felt the same way. Yep, we would pick cotton and stuff to try to make it. And when it was time for me to go back to school, I didn't want to go. And then wasn't long after that I got married. So that blows that. I got married and had eight children. One son's down in Virginia. He's a teacher. I'll be glad when he get back.

My mother started contractin' in 1947, I believe. The lady's place that we stayed on was a widow lady, but she had lots and lots of land—cotton and stuff. So after that, I think she died or sold out or something. So we moved to a little place called "The Flats" in Coolidge. We had a lot of transients come in. My mom didn't like that too hot. So she bought a place in Randolph, and she started contracting. She was a hard worker. I really felt sorry for her, she really worked hard. I think that was part of her troubles. She worked right along with the workers. When people contract, if you didn't like workin' with this bunch here or they were too rowdy or somethin', you would want to go wherever you felt comfortable. A lot of people came out here on full trucks and things from Oklahoma and Texas and different places. And then a lot of them stayed and a lot of them went back. Well, a lot of Randolph people went to California. And after a while, after the machineries came in and that made a short season, so a lot of them went to find other jobs. There's also pickin' up there in California too. It's different. You know, it's not cotton, it's fruit mostly. Round here it's mostly just the one type of work, crop work mostly. In summer we have choppin' . . . choppin' weeds. And in the fall we have pickin', and that will last for a long time. They used to go sometimes last of July and August up until January or February. That's a long season.

And way back, they wasn't so particular about kids goin' to school. But now they have to go to school. Everybody worked. My brothers, they worked in the evenings. Them ten kids would get off the bus and then the parents would have a little work and they'd work and pick cotton in the evening. But it was just hard, hard times. It was hard. And my brothers, there's two of them went to school—Melvin and my older brother. My other brother, not my oldest brother, but the one next to him, he went to college. And Melvin, he went to school a lot on his own to get where he is.

My mother wasn't a contractor like big contractors. You just sent a few people, about as much as she could handle. She wasn't, say, the contractor that you might think of like fifty or thirty, it wasn't like that. She just worked so hard and tried to save her money and did the best she could. Sometime they pay by the row, and sometime they pay you by the pound, sometime, it depends. In choppin' cotton, sometimes some farmers would pay you by the row. If you was a fast chopper, you could make money, but most of us slower type, steady people, they want to chop by the day. You know, you get paid so much a day. Three dollars, five dollars, or whatever the price was at that time. Well, that's the way it was! But see, when we came out here, bread was eighteen cents a loaf. A big loaf of bread. You can't pick cotton and afford that. You know, you can go buy flour, making this or whatever, because to go to the supermarket is impossible. Pretty soon everything you put your hand on was a dollar now. Just about everything you buy, especially in a package, is one dollar, two dollars, and cereal is out of this world.

We was all growin' up. But my brother, he was too little to work, my baby brother, Melvin. So in the summer he mostly would stay home. But like in the fall we'd get up early in the morning and you'd go 'til at night, then you'd have to come in, wash your clothes out, and hang them on the line. In the mornin' you'd grab 'em and put them back on and get out again. Then you'd come in at night and took you a little bite to eat, get your bath, and go to sleep. No TV, just radio. Kids like to play ball, the same thing. They had bikes and stuff. Just like they do now. But they may not have as much as they do now, but we had. I'd sit 'n' play ball and bat.

When we moved to Randolph, in 1947, there wasn't too many people in Randolph. Where we lived we'd see all over Randolph. But it was later that several houses went up. We bought a house. That house came with seventeen acres. That whole block where my brother lived, that street that runs north to south, all that land back that way to the railroad, she used to own all that. Back then it was all desert land, though. From Miss Selmon back comin' on, goin' south, back to another little ol' shack house right then, then back east to the railroad. That's quite a bit of land. And she had an old-fashioned pump, you know. It was an old dug well, you know, the kind you put a bucket on and you let down. You drop the bucket down in there with insects and everything. You draw it up with a long rope. It dried up. But you know, at the farm the water dried up there too. And then she had another well made, the one with the hand thing. But she wanted it done and she got it done. Well, see, back in Arkansas you had the old-fashioned well, 'cause you didn't have [laughter] . . . have all the bugs. That's the way people are. They care. They have this with them wherever they go. People everywhere do things differently. The well was right between the store and the house. It was kind of around the edge of the alley; there was an alley runnin' through there, too.

You had to bathe every day. That ol' washtub, that's what you bathe in. You get you some water. Haul your water to get your bath. When she bought the original house, it wasn't but two rooms. And then she added a kitchen and another room. She didn't get the bathroom until later. 'Cause the bathroom was outside. Boy, I tell ya, it's something these young people don't understand! You just did what you had to do with the water, goin' to school and stuff. But you know, it seemed like it was real hard. It was hard. But you didn't realize that at the time. It was just somethin' you had to do, and you did it.

My mother was very strict. The boys and people used to always howl, "That's my daughter Vera." That's how she made her kids mind. How she brought 'em up. Of course she did spank us. When we was growin' up Melvin used to tease. He's one of these guys who like to tease everybody. He wasn't mean to 'em, but just always teased you plenty. He was the youngest. He used to ride on the sack with mother. No tellin' what that he did. Yeah, that boy was somethin' else.

The Fields owned this place. They got this place. I married a Fields, his name is Samuel. We was married in 1951. I have eight children. I have four boys and four girls. Well, all of them are around here but one. And he's in West Virginia.

I worked at the Training Center from '67 to '80, then got disabled with arthritis. I worked with retarded kids. That's the only work experience I've done outside beside raising the kids. Raising kids and picking cotton, that's all.

I wanted to be a nurse, but I couldn't do it. Why, tired like I was, kids every year, it was hard. My mother was too busy working to remarry. Well, she had friends I guess, but marry, we wouldn't approve of it anyway, 'cause I guess it would have been wrong. I don't know. She was wantin' to put in a grocery store, but she never got the chance. She was pretty progressive and always was lookin' to invest in something. She was a quiet type too. She wouldn't talk, and you'd have a hard time getting a reason sometime.

She died at age sixty-two or sixty-three. I think it was arthritis that took her. I think it was the disease that just go through the family. Mama's family has heart trouble. People don't usually die from arthritis, but they do from so many other things. Well, she had a stroke. She had a stroke, but she had heart trouble too. So what do you do? That's about it. The rest is history, you know.

I assume blacks did better than Joads (check this):
(1) not cutting other residents out of (higher-pay) work so better wage, less competition, less local antipathy
(2) post-war so Depression over tho still local hard-times perhaps

V. “Turn Not Away from Their Smiles and Their Tears”

Educating Black Migrant Children

> "The whites called you names, and you ended up callin' them names, and the next thing you know, you're out there fightin'. And the teachers, they'd whip you."
> —Ron Jordan

Studies and reports from the Depression era of the 1930s to the present have documented, over and over, the plight of the underclass. The underclass also happens to be underfed, undermedicated, underhoused, underpaid, underrepresented, and undereducated. Its members have been our voiceless minority for decades; they are urban, and they are rural. When we include the dynamics of color, race, ethnicity, gender, language, and region, the problems surrounding this particular population are multiplied.

For decades, we have heard of the Appalachian poor, the Kentucky and West Virginia coal miners, the southern rural Mississippians, the Oklahoma Okies, the Arkansas Arkies, the illegal Mexican immigrants, and others who have struggled to make a better life for themselves and their families, only to die no better off than they had lived. James Agee's *Let Us Now Praise Famous Men* is a poetic account of one special group in Appalachia and its "unlovely poverty." Agee's work, as well as essays, studies, reports, and songs, deals mainly with the white poor, who are often the most prejudiced toward blacks. Seldom is there recognition of the black poor who suffer those same ills in the same places, as these groups compete for work and other sustenances. Richard Wright's proletarian work, *Twelve Million Black Voices,* forces us to hear their bitter and imploring cries.

One of the greatest and freest gifts in America is public education. Yet the chronicle of education denied is long and tragic. After slavery ended in 1865 and the promises made to ex-slaves by their former owners did not materialize, free black people were simply given another yoke to bear. With no

Courtesy Arizona Historical Society Library

A black public school near Cashion, Maricopa County, Arizona, circa 1932.

education, no "modern skills," or any sizable ownership of anything, they had to be resourceful. The open Western frontier had its lure for many, and this generation of hopefuls was told of work west of the Mississippi, and thousands of acres of cotton waiting to be picked. Little did they know that the western frontier, specifically Arizona, was still quite southern in thought and action and that segregation would follow them along the highways and back roads and through the towns and farms. This harsh but familiar experience and the extreme heat of the Southwest deserts added to their dismay.

The towns of Casa Grande, Coolidge, Eloy, Florence, and Randolph, which were surrounded by migrant labor camps and cotton farms, were suddenly faced with the charge to educate black migrant children. White teachers did not want to teach them; neither was there space for them in the public elementary schools. Mexicans who were already in the Southwest in significant numbers faced similar prejudices. The Native Americans, though not liked either, were better tolerated because of the federal funding they brought into the school systems from the Bureau of Indian Affairs.

The history of school segregation in Arizona goes back to an amendment passed by the state legislature in 1909, which said that pupils of the African race must be separated from students of the Caucasian race "except in high schools." It also said that an election could be called by 15 percent of the residents of a district if twenty-five or more Negro students were to enter a particular high school. What generally happened was that wherever a town or district saw a significant increase in the black student population, it would then build a separate school for the "African" students. Phoenix did that in 1910 when it saw that "fifty-five African Americans were among its 3204 students."[1] Wherever black students were given permission to enter certain high schools, they were put in separate classrooms. Thus, a practice statewide was to bring in a black teacher, or build a separate school, in order to separate black and white students from each other. Subsequently, black schools appeared in Tucson, Douglas, McNary, Mobile, Cashion, and Vicksburg, which all had increasing black migrant communities.

In 1934, the Southside Colored Grammar School was opened in Casa Grande and a black woman, Rebecca Dallis, became the teacher. According to historian Mary Melcher, "Dallis juggled 40–70 students in a room no bigger than today's portable classrooms. Students huddled in groups, used books with pages missing, wrote with pieces of chalk that were less than a half inch long and gathered in a smaller playground than those of white schools in the area . . . every time they went to the board to write, they were reminded of their second-class standing."[2]

The federal government and the state government of Arizona created the Borree Corner Elementary School between Eloy and Randolph, about one mile west of Coolidge, to accommodate the children of black migrants and supplied black teachers to instruct them. Borree Corner School was named for a white woman shopkeeper who had a store across the road from the school. It was part of the Coolidge School system and was operated by the state. It opened in the mid-1940s as a segregated elementary school and remained one until it closed in 1956.

Housed in a small building, Borree had two rooms; one was used for grades one to four, and the other for grades five to eight. Each room held between thirty and forty students with one teacher. Women teachers, in addition to teaching all subject areas and music, also coached the various teams, including football. Irma Warren Leos, who taught at Borree for nine years, also drove the school bus, picking up and delivering children from the surrounding communities and labor camps. She says, "I lived on the campus during the week in the housing provided for us and drove the bus home to Phoenix on the weekends and drove back to the school on Sunday afternoons . . . it was a big ole bus too!" At the end of the school day, she, like many other black teachers, would accompany the children to the cotton fields to pick. But she says, smiling, "they told me I picked too clean and too slow."[3]

After school integration was mandated in the late 1950s, some black teachers were given jobs in the Coolidge School District, "downtown." After Borree closed, its students went to Eleven Mile Corner School, about three miles west of Randolph. The interview with Matthew Batteau and his wife, Elgie, provides vivid insights into that school's operation, the educating of black migrant children, and their personal experiences with the prejudices they, too, faced, despite the fact that they were college-educated professionals. Mr. Batteau, for example, was denied adequate housing in Coolidge and commuted sixty-five miles each way from Tucson to Eleven Mile Corner for three years. These incidents further illustrate the burdens that made teaching and learning more difficult for the black migrants.[4]

According to John Yantis, "Casa Grande entered into a cooperative effort with the City of Eloy in July 1947 to build a segregated high school at Eleven Mile Corner, eleven miles from Coolidge, Eloy, and Casa Grande. . . . Because of coordinating problems with Eloy, the school board decided in May of 1949 to integrate the city's high school."[5] Even though some high schools in the state were integrated already, the elementary grades were not. Some past students remember those early years with bitterness.

In his 1978 account, "The First Negro Pupil in Casa Grande," Merrill C. Windsor cites two cases of elementary children being separated by race. One solution was to "set up a desk for the girl in the hall and have the teacher hear her recite during the recess periods." Another was to "secure and rent a neat clean tent house just across the street from the school and hire Dan Peart to teach half-day sessions." He adds, "Some of the same people who objected to integration . . . complained that it was costing half as much to educate two little Negro pupils as it cost to educate twenty-five whites." These examples demonstrate the extremes to which powerful and bigoted whites would go to exclude blacks from receiving an education.

It is a fact, however, that segregation sometimes brings out the best in us. For example, the students who attended the segregated schools in the Casa Grande Valley region speak glowingly of the wonderful teaching and learning that went on under their "Negro" teachers, who cared for them, taught them well, and became, in many instances, surrogate parents. Some teachers would, after school ended in the afternoons, change their clothing and pick cotton with the students and their parents, who were probably eager for any help that would add a few more pounds of cotton to their sacks before sunset.

The schools in Arizona were not desegregated until the mid-1950s and early 1960s. By 1962, desegregation, which was accomplished gradually, grade by grade, was complete. Of course, this is also about the time that cotton farmers adopted mechanical pickers and machines began to do most of what day laborers used to do. Segregated schools began closing; some that had been all white initiated open enrollment for blacks and others.

The Eleven Mile Corner School, for example, became a Special Education School. The students who had attended it moved either to Coolidge Elementary or to Kenilworth Elementary, located just a little northeast of Randolph. By then, many who were disillusioned, unmotivated, or "just plain tired" chose to just stop going to school. Some never went beyond the third grade. This left a generation of young people ill equipped to meet the demands of a changing workplace, technically and educationally.

Some did continue to work on farms, choppin', irrigatin', or operatin' machinery, but the data show that those jobs were primarily given to white or Mexican workers. Those were the better jobs, and the cycle of migrancy started again. In their new school settings, the transferred or transplanted students would meet new prejudices and attitudes from teachers, students, and community. No longer were black teachers there as role models or caring parents; instead, many felt the hostility of whites at every level of activity. In the dining rooms and classrooms, on the playground, or at athletic events, there was constant name-calling or taunting from their white peers. Ron Jordan, in his narrative, decries the fact that he was not expected to do well and got poor or no counseling as to the options open to him for further study or opportunities. College was never a possibility; that educational system failed him and hundreds of other students of color, including the Pima Indian students from the nearby reservations on the Gila River.

Some workers and families left for places that beckoned with work and opportunities. Many left for California, with its lush valleys, fruit farms, vineyards, and vegetable acres. But many stayed and "took life one day at a time." Some did well academically and welcomed the opportunities that integration brought by continuing into postsecondary education. Randolph boasts of football and basketball stars who joined the junior college, college, and professional ranks, like Robbie Dosty, who went to the University of Arizona and became an outstanding basketball player, and Alvin Moore, who went to Arizona State University and then played for two National Football League teams. Some, after finishing high school, found jobs in business, civil service, and teaching. Some of the grandchildren and great-grandchildren have perhaps benefited from their experiences with sanctioned segregation.

What follows is a profile of the Eleven Mile Corner School, its structure and purpose, and how transiency and migrancy affected migrant children during the years 1952–1953. In that school were three races of children—black, Mexican, and white. We must also remember that, in the world of migrancy, economics and powerlessness are overriding issues and that poverty and need know no boundaries—all suffer equally, no matter their race. Poverty can be very egalitarian. The Eleven Mile Corner School is representative of many schools like it, which were set up to serve the migrant communities in Arizona and other western states. It was sponsored by the Migrant Accommodation School Program.

Notice attached to the trailers of the contractors and posted where it might be easily seen in the camps and ranches:

> *NOTICE*
>
> CHILDREN UNDER 16 WILL NOT BE PERMITTED TO PICK DURING SCHOOL HOURS
>
> ***************
>
> United States Fair Labor Laws Prohibit the Employment of Children Under 16 Years of Age While School is in Session in This District
>
> ***************
>
> Under Arizona law a parent who keeps children under 16 out of school is guilty of a misdemeanor and may be fined up to $50.00
>
> BOYS AND GIRLS OVER 16 MAY OBTAIN A CERTIFICATE OF AGE FROM NEAREST SCHOOL PRINCIPAL IF THEY PROVIDE SUFFICIENT PROOF OF AGE
>
> ***************
>
> ARIZONA COOPERATIVE COTTON GROWERS ASSOCIATION
> 1242 E. McDowell Road Phoenix, Arizona

The Eleven Mile Corner School—A Profile

The Eleven Mile Corner School was located at a labor camp built during 1938 and 1939 by the federal government. Due to the unpopularity of labor camps, farmers preferred to have their tenants live in camps that they themselves provided. The trend changed after the labor camps met certain health standards that the farmers were unwilling to meet. After the Eleven Mile Corner Camp opened and people began to move in, the problem of education for their children arose. The camp was built in the Casa Grande High School District, not in the elementary school district, although children of both elementary and high school age lived there.

The Pinal County School superintendent was contacted, and he immediately acted to get a school for the children of this camp, since neither the Casa Grande nor the Coolidge district wanted the responsibility of educating the children of non-tax-paying transient parents. He contacted the Farm

Security Administration, which initially wanted no part in a school for public education. He appealed, arguing that the school was for the education of children of "government workers." The Farm Security Administration granted 2.15 acres of land in unorganized territory and allotted $39,000.00 for the building of a six-room school at Eleven Mile Corner. This was the first public school in the United States built with federal funds.[6]

During the school year of 1939–1940, the first classes were held in the recreation hall until the school building was completed in 1940. This school was built to accommodate the children of migratory farmworkers and was, during its first years of existence, maintained by the County School Reserve Fund. Since this school was designated as an accommodation school, its purpose was to educate the children of people who, in most cases, were constantly on the move. However, there was a section in this camp designated "The Apartment Section," occupied by families who had long-term contracts with a farmer, or were contractors or foremen.

A great deal of work has been done dealing with the problems of transiency and race and their relationships to school progress. Numerous studies have examined the Arizona schools during the period 1936 to 1940, and more recent studies continue to address the same problems nationally.

Arizona's child labor laws during the height of the migrant labor activity did not permit children to work during school hours unless they were at least sixteen years of age or had finished the eighth grade. There was a time when a child who had not finished the eighth grade, or who had not reached the age of sixteen, could get a certificate to work under certain conditions, but this practice was not encouraged. We must remember that pickers worked in family groups or with other relatives or workers who migrated from the same states and towns. Child labor because of family need was a prevalent practice.

Children would be found working in the cotton fields who had lied about their ages, or whose parents had kept them hidden in the fields. Often these children were not allowed to come to places where they could be seen except at quitting time. The children would pick cotton, put it in their parents' sacks, and then hide in the field while their parents went to the weighing truck with the cotton. It was common for parents who were not much concerned about their children's education to lie about their ages, because the extra hands in the field would pick more pounds of cotton on a daily or weekly basis. Farmers became more careful about enforcing the child labor laws over time, as charges were filed against those who allowed schoolchildren to work in their fields during school hours. At one time a three-hundred-dollar fine was given to any farmer with an underage child in his field. Parents would be fined fifty dollars for keeping a child under sixteen from school.

The federal government intervened, and most of the contractors who saw to it that the cotton got picked put signs on their trailers plainly stating that "no children under sixteen years of age will be allowed to pick cotton during school hours." Again, some families got around this law by lying about the ages of their children. Some farmers, to prevent charges for violation of the Arizona child labor laws, required birth certificates when there was any doubt as to the age of a child wishing to work in the field. From the state's point of view, when children attended class regularly, more money was allowed for the maintenance of the schools through county and state funds.

The attendance at Borree Corner School and Eleven Mile Corner School for blacks was in some respects about the same as that in other schools, except for the fact that some parents did find excuses to keep their children out of school. Sometimes children who were absent due to a reported illness were found keeping a smaller brother or sister. The most prevalent reasons, other than illness, given by pupils at the Eleven Mile Corner School were to keep a younger brother or sister who was not of school age and to help do the family washing. It was a practice with most families when they first moved into the camp to keep their children home the first day. When asked why they did not come to school the first day, the answer in most cases was, "I didn't have any clean clothes; we had to wash," or "We been travelin'." It was true that in some of the camps the facilities for washing as well as for bathing were very inadequate. In the case of Randolph, for example, water had to be hauled in and paid for. In many instances a farmer did not provide this service for the workers and pickers.

Often, parents left the camps early in order to get to the fields, leaving the duties of getting ready for school to their children. Where there were many children in a family, the responsibility for getting younger brothers or sisters ready for school was given to the older ones. According to Batteau's study, "The fact that they probably did not sleep in a comfortable bed, did not have the proper food for breakfast, or did not clean up properly, causes them to be sluggish and irritable. As a result they are not able to think properly and cannot retain much of what is taught during that day at school."[7]

The cabins and apartments in the camp were numbered; when the children registered, the numbers of their dwellings were recorded so that the teachers would know how to contact them or their parents in case of an absence or emergency. Often when a child did not show up at school a custodian, who also took care of the duties of attendance officer, was given the cabin or apartment

Migrant housing.

number and immediately checked to see why the child was absent. The school was conveniently located near the camp area where the children lived, and such checks could be made in a few minutes by walking from the school to the home of the absent child.

There was no set procedure for dropping children from the school rolls. When a family moved, other children attending school would report this fact to their teacher, or the teacher would ask if a certain family had moved. The teacher usually learned such information from other children.

When the children knew in advance that they were going to move, they would report this to the teacher. The teacher would then issue a transfer showing the grade, the progress made in that grade, and the dates of entry and withdrawal. "There have been occasions when a child may enroll in school, pay for his lunch for that day and before noon the parents have decided to move. On some occasions they come to the school all packed up to get their child or children and get on their way to another camp."[8] The child would be then given a refund of his lunch money before leaving.

The teachers, especially where there were two or more grades in a room, as at Borree School, usually had multiple things to do, which interfered with instruction. In addition, late entrance, early withdrawal, and irregular attendance were quite common where migrant pupils were enrolled. Often students would change schools two to six times in one year. One report shows one student attended nine schools in one year, another ten.[9] The greatest percentage of transient migrancy was among whites, followed by Mexicans. Black students and families, although transient, showed longer tenancy and more consistent school attendance.

It appears that race, then, is an indicator of migrant labor transiency in the Casa Grande Valley and that access to work and mobility was easier for whites than for blacks or Mexicans. Originally, Mexicans faced not only ethnic and cultural biases but the barrier of language as well. During the school year 1951–1952, 100 percent of the Mexican pupils transferred during the first grade; there were no permanent pupils. Eighty-four percent transferred in the second grade, and only 16 percent were permanent. In the third grade, 76 percent transferred and 23 percent remained permanent. In the fourth grade, 84 percent transferred and 5 percent remained permanent. There were no permanent Mexican pupils in the sixth grade.[10]

Among racial groups, white pupils appear to have made more rapid progress than blacks or Mexicans. This may be due to the fact that, in some sections of the country, white children had better schools, facilities, and opportunities provided for them by the dominant society before migrancy began.

Studies also show that, by gender, migrant girls did not do any better than boys; the dropout rate for girls was often higher as expectations that they would help to care for siblings and the home grew. Girls also got married earlier, often after meeting their prospective husbands at school or in the fields. But once schooling was over, they also struggled within the system to raise another generation who would enter a segregated society like the one in which they grew up. Some became quite bitter about not getting a better share of America's promise and, seeing the futility, simply gave up and left school forever.

In a rebuttal to an article in the *Arizona Republican* in June 1916, which said that children had suffered from ostracism that they would have been spared if they were segregated, William P. Crump wrote, "I wish to deny emphatically and I am aware that the proponents of this retrogressive move would like to create such an impression. There may be a stray colored person or two who have drifted in from Texas or Arkansas who may have an idea that they would prefer separate schools. This is because they do not know any better. . . . They do not comprehend, nor will they ever comprehend the present and cumulative disadvantages and injustices of such an arrangement."[11] When a white woman who had been a student at the newly desegregated Kenilworth School was asked what she remembered most about that time, she said, "Oh honey, we wouldn't win any games without 'em." These two responses have little correspondence. Crump, despite the fact that blacks were admitted for other purposes, sees integration as a disadvantage, generally because of the problems of color and class lines inherent in it. But the fact remains that students, no matter why they were separated from white teachers and white schools, speak glowingly of their experiences with black teachers in black schools. For those teachers and their students it was also a very special time and experience, a unique period in their lives. The cycle of migrancy, and the transient lifestyle, did not deter some from putting down roots or settling, so as to provide a better and more thorough education for their children and themselves. Some went back to school to get high school diplomas; some went to college and received degrees; some never forgot the opportunities denied. Migrant children know what they feel and have lived the results of both systems.

The renovated Southside Colored Grammar School, on the grounds of the Casa Grande Historical Society.

Migrant children.

Courtesy Jordan family

Matthew and Elgie Batteau

Elgie and Matthew Batteau both taught in the Migrant Accommodation Schools at Eleven Mile Corner. In this interview, both Matthew and Elgie Batteau talk primarily about their lives as teachers and the prejudices they encountered at every level, from getting their education to giving others an education. They also talk about the joys of their vocation and of helping those who "had such poor chances." Their amazing careers totaled sixty-three years of teaching!

Matthew Batteau was born in Houston, Texas, in 1910, the sixteenth of seventeen children. He jokingly tells of how people would think he was a Louisiana Creole of French origins because of his last name, noting that the history of "Negro" Americans shows such great misconceptions, because naming comes from the white families who owned us. Mr. Batteau earned his B.S. degree in education from Houston College for Negroes in 1941. This college was renamed Texas Southern and is still predominantly black. He is also proud of the fact that he was president of his class for four years. He says, "My first year at Houston College for Negroes, I felt inferior to the students who had gone to day school, and I didn't go to any freshman class meetings." As a teacher, he understood how a school building would be intimidating for many students and parents. He taught at the Eleven Mile Corner School for ten years, then in the Tucson Unified School District for another seventeen years in the Special Education program, retiring from there in 1975 after twenty-seven years in education.

His M.A. thesis, "Child Accounting in the Eleven Mile Corner School," was based on his teaching and involvement with the Eleven Mile Corner Road School. In 1941 he earned the M.A. degree from the University of Arizona. He vividly remembers events, students, and the various ups and downs he and his wife, Elgie, and their daughter, Flavia, experienced up there "in the cotton."

Elgie Mike Batteau was born in Victoria, Texas, in 1905 and came of age there. She was valedictorian of her class and received her formal graduation at Paul Quinn College in Texas and at the University of Arizona, where she earned a B.A. and an M.A. in education. She taught at the Phoenix Union Colored High School for nine years, during which she worked to have the school renamed Carver High because she hated the word colored as a signifier. In the 1960s, the school was renamed the Phoenix Union High School. In the 1980s, the Carver name was revived, and the building now houses the Arizona African American Museum.

Mrs. Batteau also taught at Eleven Mile Corner School for seven years. Her daughter, Flavia, attended the school for four of those years and learned the value of hard work, sacrifice, and empathy for those less fortunate than she. She retired from the Tucson Unified School District in 1970 after thirty-six years of teaching.

The Batteaus married in 1941 and had been in Tucson for most of the years since then. I knew Mrs. Batteau and her daughter for a number of years as my sorority sisters yet never knew the story of their professional journeys. It was one of the Batteaus' students, Ron Jordan of Randolph, who told me that they were his teachers at Eleven Mile Corner School and what caring and wonderful parents and teachers they were to all. It was an amazing connection for me, and thus their involvement in this "history lesson."

Matthew Batteau was the more fluid and energized during the interviews, and his part of the story flowed so well that I said very little initially. Elgie Batteau spoke sparingly until I began to interject questions directed to her. She is a gracious and refined woman with a strong sense of who she is and where she came from, and she is seemingly fearless of those "in power." She was not well when we visited last, and her various illnesses, including failing eyesight, took her life in April 1994. Mr. Batteau says "the burnin' of the cotton bolls kind of affected her health." But both say, "Those children were eager to learn out there, and it was a joy teachin' those children." Speaking of the transient nature of migrant children's lives, they commented: "They may go someplace else to harvest, pick cotton, harvest vegetables, go into California. They could go all the way up to Washington and pick apples. Those kids knew more geography than we did."

These two devoted teachers also took their cotton pickin' clothes to school with them and would get on the truck after school and go out with the kids to pick. They said, "It did something to the kids to let them know that it wasn't any disgrace to pick cotton. But we also taught them to try to learn, where you won't have to do this the rest of your life." Many of them still remember that, and the Batteaus.

M: My high school work was done in night school. I went to two high schools to get enough credits to take a college entrance exam, which I did, and passed. Then I entered Texas Southern University, but then it was called Houston College for Negroes.

My first year at Houston College for Negroes, I felt inferior to the students who had gone to day school, and I didn't go to any freshman class meetings. So about the end of the school year the adviser of the freshman class sent for me and told me they were having a meeting, and he wanted me to attend the meeting. He told me where it was and what time and everything, and I went to the meeting, and at that meeting they elected me president of the sophomore class. And thereafter, I was elected each year as a president of the junior class and the senior class. When my class graduated, there were seventy-five of us in the class, and I marched right behind the faculty with my class right behind me, which was an event to remember, and after graduation in 1941 I came to Arizona. I was working for Sears, Roebuck in Texas. I had been there for some ten years, and I had a little stock with the company, and that was the only money that I had to go on so I drew out my stock.

There wasn't a chance of getting transferred to the Sears store in Tucson because it was a small store. I had worked in the rug and linoleum department in Houston. I put the samples on the floor, unloaded boxcar loads of rugs and linoleum by myself—whole boxcar loads! When I went back to Houston I gave them two weeks' notice that I was going to quit and come to Arizona. I was only getting eighteen dollars a week working sixty and sixty-five hours a week plus going back on Sundays a lot of times, working half a day and a lot of nights. When I wasn't going to school, I would have to work cleaning up when some of the big shots from Chicago were coming to inspect the store, for which we got no pay whatsoever.

I would say that eighteen dollars was low pay scale, but we were workin' for a fella that was a slave driver. His name was Ed Skroggins, and he got all he could get out of us for nothin', and that's exactly what he did. I would grit my teeth a lot of days, and I said, well if I get an education, I won't have to do this the rest of my life, and that's what I pursued, an education. I didn't have a job so I would take the first job I could get. I worked at the Santa Rita Hotel as a waiter for a while. Then I got another job working as a waiter at the Circle Z Dude Ranch just out of Nogales, Arizona. During that time I waited on some very influential people. I waited on the vice president of the William Wrigley Company and the president of Hammermill Paper Company, and those two men kept in touch with me the whole time I was in military service, but after I got out of the service somehow we lost contact. Prior to going into military service I worked as a stock clerk at Davis-Monthan Airbase. After getting discharged from the military, I went back to work at the airbase until I decided to go back to the University of Arizona to qualify for teaching in Arizona.

When I came back from the service I applied for work here in Tucson, and I wasn't able to get on here in Tucson even after I had gone to school and qualified for elementary teaching. The only possibility for blacks here in Tucson was elementary school. They couldn't teach in the high schools. Only whites could teach in the high school. So I applied at the University Placement Center, and they heard about this job in Casa Grande, so I went to the superintendent. Lewis Taylor was county superintendent of Pinal Schools at the time. I applied and got the job as head teacher of a two-teacher high school at Eleven Mile Corner. That was to teach children that rode the bus from Casa Grande and Eloy, because they were not permitted to attend high school in Casa Grande or Eloy. That only lasted for one year, and after that I went into elementary on the camp of the school that my wife had started for the children. I had grades four to eight for two years, and then after we integrated, we went down to the white school at Eleven Mile Corner. I taught grades four and five for two or three years there. The last year I taught grades five and six, but I had experience teaching from the fourth grade through the twelfth grade.

I think I did my certification in one summer. I had to go to both sessions 'cause I only had to take elementary tests and measurements, and Arizona

Constitution. Of course, all teachers have to take that Arizona Constitution course. I qualified in one summer for elementary. At that time, Mrs. Batteau and I had corresponded for nine years, and I came out here to see about getting work since we had been going together for so long; well we were thinkin' that maybe we'd get married. I wanted to get my education first, so I would be on the same level with her.

That two-teacher high school only lasted one year. Then I went into elementary along with my wife there on the camp where we were teaching. We taught on a stage of this auditorium. We just had a partial partition between her class and my class, and when we would get an overload of students, we'd have another teacher come in for part-time.

Just as you go in the camp, there was the office buildin' over on this side, and right in front of the office buildin' was this huge auditorium. I think in back of the auditorium was the washroom. They had stationary washtubs and everything, and I think maybe they did have dryers for the cotton pickers, where they could do their laundry and everything. They had all modern conveniences, but it was twenty-five-cycle electricity. The government built this farm labor supply center in the 1930s, and that's where we were. It wasn't a school auditorium, it was an auditorium for the cotton pickers. They had some people that lived there permanently. The ones that lived in the apartments stayed the year round. But the ones that lived in the Quonset huts, they were just seasonal.

At school closings, we would have programs for the people, and the people were just surprised that we were doing so much with those kids and so much for them too. And we would solicit clothing and would take trunkloads of clothes down almost every two weeks at least. Trunkloads, just trunkloads of clothes for the grownups and for the children too. The children would come from the southern states. They had some from Mississippi and Texas, New Mexico, Oklahoma, and I believe we had some from Arkansas too, but most of the children would come from those states. We had one contractor that lived in Oklahoma, and he knew a lot of people back there, and he would go back to Oklahoma—Mr. Henderson.

And then there was another family that lived down there. I think one of the boys lives at Randolph now. One was Lems, and one of the boys I talked with, Lee Andrew Lems, got to be a United States marshal. There's another family that their boy made me some lamps. I have one of the lamps in my bedroom, right now; he made it in shop down there. His father died first and then his mother died. There was a bunch of them. Then one of his sisters died. I taught all those kids, and I can't think of their last names. It's a long time ago. This one boy lives in Randolph because I think I have his telephone number down there. I have forgotten his first name, but Rushing is his last name. I think it's Clayton Rushing.

Tell me what a typical teaching experience was like at Eleven Mile Corner School—what materials were used, where they came from, how the kids responded, the size of the classes, those kinds of things.

Well, I had no trouble with those kids out there as far as having more than one class in a room; you just had to do the best you could, divide the time the best you could and do what you could. One thing I tried to teach the children was that they would have to either be studying their lesson or be quiet because they couldn't disturb the other children. They were responsive like that, they would obey. They were far different from the children after I came here to Tucson. Far different. Those children were eager to learn out there, and it was a joy teachin' those children. My wife said many times that she wishes that she could have just stayed out there until she retired, because she enjoyed it so much out there, but the burnin' of the cotton bolls kind of affected her health.

We would get the materials and books from the county superintendent. And to decide when to move the kids to another grade, well, we would give them tests after they would complete their course. For instance, if they were in the fifth grade when they would do the work that was required of them we would pass them to the next grade. And some of those kids we taught out there went to Phoenix and went through the schools in Phoenix and got very high marks. Some of those kids, I think, are teaching now.

There was an incident once with a Mississippi family. There was a boy that I had taught, and his parents took him back to Mississippi and he was very unhappy there, and they finally came back and they said that the teachers there wanted to know who had taught him. I guess they told him that a nigger teacher had taught him. I'm sure that's the language they used. But anyway, the boy was very unhappy there, and he came back and he was a good kid. I had no problems with him in spite of the fact that he had heard all of this from his parents and other people too. He really liked what I had done for him, and his parents did too.

What about your wife's involvement with teaching? You had mentioned that she was up there for seven years. Your daughter was up there also. Tell me about that whole era.

Oh, yes. Well, the first year we went up there our daughter, Flavia, was not old enough for school, and we left her in Tucson with her grandmother. We would only see her when we would come home weekends. But the second year we decided that we would take her up there, and she would come over to walk her mother home 'cause we didn't live far from the auditorium. It was less than two blocks where we lived, a little place. My wife would go home for lunch, every day, so our daughter would come over and walk back home with her mother. She was over there one day when the superintendent, Mr. Taylor, came out and saw her at the blackboard writing. So he questioned whose child it was, and my wife told him it was her daughter, and he said, "Does she go to school?" And she said, "No, she is not old enough." So he said, "Well, you baby-sittin' these others, put her age up and put her in school." So that is what she did. So she graduated a couple of years ahead of herself, but she made it alright. She was the youngest student that ever went to Mills College, at age sixteen. That's in Oakland, California. Did her four years' work in three and a half years. When she came here to go to the fifth grade, I would have taught her had she been down there at Eleven Mile Corner because I was teaching the fifth grade. But she was glad that she didn't have another Batteau for a teacher.

How did she get along with the kids up there since she was a teacher's child, as opposed to being the child of a migrant worker?

Well, her best friend up there was a little white girl, believe it or not, Donna Short, but she got along very well, of course, in playing. There was one kid up there, she was a very neglected child. They were swinging. Hattie, a black girl, hit the swing, and it hit Flavia in the forehead, so we had to take her to the doctor. But she didn't do it intentionally, did she?

E: No. She didn't do it intentionally. The thing of it was, they were out there, well, she did it intentionally too. Because she was so mistreated herself that she didn't know anything but mistreatment. Just had to strike back at someone. And she knew that Flavia was our child. She took this metal seat and held it back and threw it and struck Flavia. In the meantime a little white boy, another one, had thrown a two-by-four or something, and had cut one of the kids. I happened to be in the yard that particular day, and there was Flavia with a knot on her head from the swing and there was this little boy bleeding from the two-by-four wood. I didn't think anything about it so I just went to the little boy because he was bleeding so. And the teachers were all standing there saying, "How could you leave your child with that big knot on her head and go to somebody else's child." But see, I didn't think about that. I saw the blood, and the kid was playing and naturally he was bleeding quite a bit. I took them both to the doctor, and the doctor told me that Flavia was hurt worse than the little boy. I didn't know that at the time. She was picking cotton out there too. Flavia picked cotton with them, but she'd get way down someplace and say, "Get me out of here."

M: There used to be a store, the Wallace's store, on the road from Casa

Grande to Coolidge Road. They moved, but I even taught their children too, and so did my wife. I had one boy named Barry and the oldest boy, I forget his name, but my mother-in-law used to do his laundry. She used to do it particularly for kids at the University of Arizona. He came to Tucson to attend the university, and now he's a veterinarian. They were nice kids.

What was the attendance like during cotton picking season?

M: It fluctuated so; it depended on when they would go and bring in a bunch of cotton pickers. You see, they bring in these children with them. Well, usually when they bring in a bunch of cotton pickers the school enrollment would go up because you would get a bunch of children and you would have to enroll them.

E: You see, that was the experience that I had with the classes that I started. When I first started I had about thirty-six children that were just roaming around on the camp and were not going to school. The people that owned the cotton were afraid to bring them in the field because they would have to pay three hundred dollars for every child they found in the field that was of school age. They were afraid to have them come in, so they just roamed around the camp. We'd come home on Friday afternoon, and when we went back Sunday afternoon or Monday morning, I'd have seventy-three children in the class! We had to get extra teachers. We had Miss Podrill, Miss Thompson, and Mrs. Lewis. And also we had that other woman, Mrs. Henderson, who wrote the nasty note and sent it to school . . . I don't remember her first name . . . she was out of Oklahoma, and now the family lives in Phoenix.

But we'd have to hire extra teachers during the peak season, which may not last over three or four weeks. They may go someplace else to harvest, pick cotton, harvest vegetables, go into California. They could go all the way up to Washington and pick apples. Those kids knew more geography than we did. And they would try to go to school in different places. Maybe they would be in school a week, wherever they would go, because the contractors were afraid to take them into the fields or into the crops because of the state child labor laws.

M: I think the children would pick after schooltime. We used to take our cotton pickin' clothes to school with us, and we would get on the truck and go out with the kids. And it did something to the kids to let them know that it wasn't any disgrace to pick cotton. But we also taught them to try to learn, where you won't have to do this the rest of your life. And we also told them, and they would live to see it too, that cotton pickin' machines are gonna take your jobs away from ya, and they have lived to see that, you see.

E: We went in early in the morning around seven-thirty, and around two-thirty, when we finished, we went in the fields. The contractors would come to pick up the children and the teachers. See, that was lawful. They had put in a school day, but the superintendent, Mr. Taylor, would come and check.

M: He used to ride his horse.

E: He would take his horse and put it in the trailer behind his car and have it saddled and everything, and go to the fields, and get on his horse to see if there were any children in the fields. It hadn't always been that way though.

What was Mr. Taylor like?

M: He was white. He was the Pinal county superintendent. He was the one I had to go to to get my job, until my wife started school, I think it was the fall of 1948 or '49. My wife went to the superintendent and told him that there's kids runnin' around that should be in school, and he said they didn't have any money. Then he decided that maybe he'd better do something about it, and he started a school and she worked for a minimum salary, but she started the school anyway. So those kids had a chance of goin' to school, and some of those same kids that went to school there, as we have told you previously, became teachers in the schools in Phoenix.

After my wife came back to Tucson one year, rather than live on a camp, I decided I would get an apartment in Coolidge, and that's when the white people would throw things. We had a Buick. It was a Century, Buick. At Eleven Mile Corner I couldn't get an apartment over where they had the apartments, you know, near the school, because I was black. I couldn't get one there, so I got an apartment, a little shack, in Coolidge, and that's where they were jealous of me. I had to finally get a gun to keep them from messin' with the car. I had running water there. It had an indoor toilet and running water. I think there was a shower. I don't think I had a bathtub. All together, I taught ten years in that area. Ten years! I taught three years after she came to Tucson. The last year I taught, I commuted. I drove sixty-five miles in the morning and sixty-five miles back home in the evenings, five days a week. Sixty-five miles!

Well, that year I got work in Tucson in the Special Education school. I went out to cook at a camp for retarded children. We went to a place on Silverbell Road and had to clean up that place, and it had closed down as a dude ranch, and they left cans of grease where big rats had drowned in the grease and was still there. We had to clean up all that filth in order to get the place ready, and I was going to cook for them, which I did for three years. We cooked one year at Silverbell, and my wife and Flavia had the crafts department, and then we went to Sahuaro Vista Ranch. We were there one summer, and then the last summer I said, "This is too much—work all day with these kids and then go with them to some ranch." So we went to the Forty-niners Ranch, you know, that's one of the big dude ranches.

Mrs. Batteau, I am waiting for you to tell me the story about the naming of Carver High School, how you got to Phoenix and how you lobbied for that.

E: Mine isn't as interesting as his. But when I started teaching they had one school here in Tucson where black teachers could teach. Now, when I say black, what are we now? [Laughter]

African Americans.

E: When I was writing my thesis, I wrote about the Negro as a soldier from the first American Revolution through the Spanish-American War. Anyway, when I said I don't know what we are, because when we first started off and I started looking at congressional papers and all, it had n-e-g-r-o with a small "n." And then we went to a capital "N" Negro, N-e-g-r-o. Then we were, I can't pronounce it the way they pronounced it, "cullud." Colored but spelled with a "u" . . . "colored" folks—we went to that, yes. We went from colored to black.

M: No, Negro first, with a big "N."

E: Yeah. And then we went to black and then we, now we're African Americans. So I don't know what we are. And I was told when I first applied that they had no vacancies at Dunbar. And I told Mr. Rose, he was the superintendent then, I said, "Well, Mr. Rose, with all due respect to you, I am not naming any special school to go to. Whatever school there is a vacancy, I would like to feel that I might be able to get work there and all." He says, "Well you know you can't work anyplace but Dunbar." Because that was the black school at that time, and from the first through the junior high they bused the kids to Dunbar School. He finally told me that I wouldn't have to come to the office anymore, and I said, "Well I want you to know that there is a (I used the term *Negro*) a Negro teacher looking for a job in the schools here in Tucson in the Tucson Unified School District." So I think they finally got tired of seeing me and they finally gave me a position at Dunbar. Which was the only place I could get work. He tried to tell me why didn't I go to the Amphi district when he *knew* they were *not* hiring black teachers there. Even at that particular time I don't think they had any black students there either.

M: I think Mrs. Hale was one of the first black teachers there.

E: Anyway they hired me, I think, to get rid of me because I was letting him know that I was interested and since there was just one place for me, I was interested in getting a position in that one place. I happened to have been living in the Steinfeld home, where I lived for the two years when I was attending the University of Arizona. The Steinfelds [wealthy white Tucson merchants] knew all of these people and so on.

Were you boarding at the Steinfelds?

E: No, my aunt was a housekeeper there, and we had an apartment in the house. We had two bedrooms and a bath and a sitting room on one side of the house, and the house was built in a "U" shape, and the Steinfelds lived on the other side. Anyway, I was lucky enough to get work. Worked at Dunbar for four years, but I was still wanting to go into high school because I had had high school training rather than elementary training. And they had this colored high school in Phoenix, Phoenix Union Colored High School, it was never just Phoenix Union, it was always the colored high school, and I taught there. But my second year there, I asked the principal to take me to the superintendent's office because I wanted to talk with him. That is when I asked him whether they had a name for the school. He told me, "Yes." I said, "But it's always referred to as a colored high school. Couldn't it have some other name other than that?" And he said, "Well, what would you suggest?" And I suggested West High School, West Phoenix. And he said, "That's on the drawing board." I said, "Well, South Phoenix." He said, "That's on the drawing board." And I said, "Well, what isn't on the drawing board then?" And he said, "Well do you know some outstanding Negro that you would like to see the school named for and everything?" We had been using a little black history booklet, in fact it was the only thing we could find at that time that could be used in the school. I told him we were studying about George Washington Carver. He said, "Well, do you want to name it Carver High School?" And I said, "Yes. I would rather say I was teaching at Carver than to say I was teaching at the Colored High School!"

I left that school during my ninth year. I left in November, and Flavia came in April; I was embarrassed to stay. The superintendent told me I would have to stay until I found someone to take my place. And I didn't know anybody that I thought would want to break into a new year after school had started and everything, but I had a friend. We were in class together at the university, and I asked her if she would teach. She said, yeah. She took my place, and Flavia was born in April, and my friend's baby was born in August. So, that was the end of my stay in Phoenix. Then I came here to Tucson as a permanent substitute. They gave me a full salary but no contract. I taught day by day, but I taught the whole nine months. I was told I would have a nine-month job but without a contract, but my salary was more than some of the teachers that were teaching, although I was teaching as a substitute.

And at that time they weren't supposed to be hiring married women. In Phoenix, even if it wasn't a state law, it was just a school board ruling. I taught seven years longer. When I went to the superintendent in Phoenix and told him that I had just built a home here for my mother and it had to be paid for and I had no job, and I couldn't afford to lose what little money I had already put into it, he says, "Well, because of the war situation we'll rehire you as a married teacher—married woman—see, but, I'll give you a contract this year, but next year you will not get a contract because we just don't hire married women," and that's when I asked if they had a law against it, and he says no, it's just a ruling. Each year he would say, "Do you want to come back?" I would say, "Yes," until I found out that Flavia was more important to me than the contract.

What made you get involved with Mr. Batteau's school in Eleven Mile Corner?

E: I came there because Flavia would have to go on the bus to all these different farms, and sometimes it was seven-thirty or eight o'clock at night before the buses would get the children home.

Was that because she was black?

M: No, the whites and all rode the same bus, but they had so many farms to go to to pick up children for the high school.

E: And so many miles to cover. She would be getting home so late at night and, see, in the fall it's dark at seven. She would still be on a bus tryin' to get home. And I visualized everything happening on the bus, everything, you know. As African Americans or as blacks, whatever we are, you know, we never know what's going to happen, and she would be the only little black one on the bus coming to Eleven Mile Corner. I was just afraid of what might happen. That is when I moved here with her.

She still remembers when she was in school at Eleven Mile Corner. She enjoyed it. She liked it there, but she didn't want to go to another Batteau as a teacher! [Laughter] We didn't want her to feel that she had any privileges the other children didn't have because she was our child. I think sometimes we were a little bit harder on her than we should have been so the other children wouldn't even think it, you know. I know one little girl came up to her one day and slapped her. Flavia started cryin' and the girl's print was all on her face. The handprint. And I asked her, I said, "Well why did you want to slap her?" And she said, "She thinks she's white." She was light skinned, see. I said, "I don't think she thinks that because she knows what she is." This was a very dark little girl. Anyway, I said, "Well you know that the practice is, if she slaps you, then you slap her. I'll stand here and let you slap her, see you slap her." I said, "One slap deserves another if a person has to slap." So she says, "Well she's not gonna slap me!" I said, "Well did you slap her?" She said, "Yes." I say, "You see all of your handprint is on her cheek there." I said, "Well now it's her time to slap you." Flavia said, "I don't want to slap her." I said, "But you're going to slap her because she slapped you." And Flavia just kind of tapped her. I said, "No, I didn't say just tap her, I said slap her." And I said, "She slapped you first, now you slap her." So Flavia slapped her. And I said, "Now that's a ruling. If anybody slaps or strikes anybody else, the other individual gets a chance too, because you want to play ticktacktoe, play the same way, see." And I never had any more trouble with slapping! [Laughter]

Whenever I meet people from Randolph and other places who remember either of you, they speak highly of you.

M: Well, we knew that those kids were working at a disadvantage. We tried to do all we could. We talked to them and tell them positive things. We even went on the field with them and picked cotton with them to let them know that it was no disgrace to do that. The parents thought it was just wonderful of us to do that.

Do you remember a farm called the Casa Grande Valley Farms? It's outside Coolidge. Did your students come from that far?

E: That's where they raised the vegetables and things, you know, lettuce, onions, broccoli, and so forth.

M: Well, I don't know. We had some kids that came from Randolph. There were those Musgroves, they lived some place out there.

E: From Casa Grande it would be more of them that would come from the vegetable farms. Coolidge and Eloy too.

M: You see, the kids were bused from Eloy and Casa Grande. I can't remember now if the same bus picked up the kids in Eloy and Casa Grande. If the same bus picked up those kids, especially the high school kids, it only lasted for one year.

E: I think so. Eloy and Casa Grande, because it would be on a straight shot. They were using that building at Eleven Mile Corner. I don't know whether they have changed it or not, but they used the white school as a special school for special students now called Mary O'Brien Special Education School.

M: It is now Mary C. O'Brien School, Pinal, Special Education. See, Ms. LeSeur gave me this picture of the school, and this is of the school too. You went in like this, I believe [pointing at a photograph]. My room was to the right, and yours was to the left, wasn't it?

E: And it had this long porch.

What kind of pay did you make at that time?

M: I have one of my contracts here. I think I have it in one of the drawers. But I think when I first started up there, I was making $2,400 a school year. Something like that. The second year, when I started in elementary, it was an accommodation school. From one year to the other, they didn't know whether there'd be a school the next year. We worked under those conditions. But every year we would get our contract to come back for another year. I don't know if it is still an accommodation school or not.

What students remain vivid for you from Eleven Mile Corner?

M: Well, the Rushing boy, the one that made the lamps. I was very fond of him. He was a good student, and he had a sister—she's dead now. I think her name was Louise. Leonard got into some kind of trouble, and I think he went to prison or somethin' for a while, but the Rushing family, there was quite a family there. Their father died. The mother kept them out there and would work them and send them to school. She appreciated what we were doin' for the kids. Then there was a family from Oklahoma, the Ashers. Three of those girls, they were brilliant girls. Quintella, Dorothy, and June, there was another . . . Dorita, I thought. I didn't teach June.

E: June was with Flavia.

M: Yeah. June was with her. But the Asher family; they had three brilliant girls. They finished high school in Phoenix and went to Arizona State, graduated, and taught. Some of them might still be teaching there in Phoenix. I haven't seen but one of the brothers. Charles Asher came down; Freeman used to like Flavia.

Then there was Fernando Luna and Mosalina Rodríguez. Now those were two brilliant kids, too. Spanish kids. Mosalina Rodríguez was this girl's name. She was smart. Then, of course, there were the Jordan children, Ron being the one that stood out. But there were some very smart kids up there. Those kids were eager to learn.

E: They were so respectful of teachers. They admired you as a teacher, and they felt that we were trying to help them. At least they made us feel that way. They felt that we were trying to help them and they were working with us. They were cooperating, and so on. We didn't have much trouble. You would have thought that it would have been a lot of racial trouble with 'em traveling from all sections of the place, but even on the camps, they didn't find the whites calling blacks anything out of their names. In fact one Mexican girl called one of the girls black. They were on their way to the store, and the girl jumped on her. Beat her up. And the man who was the superintendent of the camp went to say something about it. He said that the nigger girl had jumped on her and everything. He said, "Well why?" And, "What did you do to her?" Said, "She did nothing." Well the girl told him she called her a nigger or something, then the trouble started. He said, "Well, you won't do that again, will you? Well, I didn't think you would! Because she stopped you."

Did you tell me that the white children went to a different accommodation school or the regular school?

M: On the camp, the white children went down to it before integration. When we left there, it still was a regular school. It became a special school since we left there. So the white children went to that school. The black kids were in the auditorium. It must have been the first year, then I think I taught two years with you [Elgie] in the upper grades in elementary, and then we integrated. I think that's the way it was. One year with the two-teacher high school, and then two years I was in elementary before we integrated to the white school. But my wife was going down to the white school teaching in the evenings, teaching at the black school in the morning and then in the afternoon going down to the white school to teach before we integrated.

Have you ever been to a camp, to see what life was like for migrant children?

M: The camp out there was the only camp that I occasionally would have to go to. They would have, I think they had, mostly hot plates.

E: They had gas. Natural gas. They had electricity and gas.

M: Yes, they had gas for cooking. And it seemed to me that these huts had a two-burner gadget that burned gas.

What about water?

M: I think there was a well where they all had to get the water. I believe that's the way it was. And then they had a washhouse where they had stationary washtubs.

E: And irons too.

M: Yes. That was a separate building. The washhouse they called it.

What about bathrooms and sanitary facilities?

E: I think they had central baths and showers.

provided

M: But at one time the little house that we lived in, and I would like to be able to show you that, it was a little brick building, but the whole thing wasn't as big as anything. It was separated. One side was the men and one side was the women and the floors went like this with the drain. My wife made the curtains and put them up. And when you pull them down the curtains would hang because, you see, when it goes back up then the curtain would be up against the glass, you know. But my wife made it look good. Had it quite livable down there. And they would keep it for us during the summer, and we would go back in the winter and go back a day or so before school and clean it up. We would put a Green Light insecticide down from the doors. One year a scorpion got on her back, crawled through that Green Light and got as far as the bedroom, got under the bed, and when we went there, the mama scorpion and the baby were petrified under the bed, just petrified—they had crawled through that insecticide. We saw plenty of scorpions down there, black widows and all.

E: We had a little electrical hot plate that we cooked on. I think we still have it in the storeroom. We got a little oven. We bought our first refrigerator for Eleven Mile Corner there on Congress Street.

Why did you write your M.A. thesis on Eleven Mile Corner School?

M: Well, we were tryin' to find the different schools that kids went to within a year, you know, a school year. That was one of the things that we were looking at. And that some of those kids survived after having gone to so many different schools during the school year. Yeah. That was it.

Ron Jordan

Ron Jordan's mother, Myrtle Jordan, encouraged me to talk to Ron because, as she said, "He's the talker in the family, he knows a lot." Because my original intention was to interview the older residents of Randolph, I was a bit skeptical. However, Ron turned out to be the most spontaneous member of the Jordan family and a great resource for information pertaining to Randolph, other residents, the surrounding farms, landmarks, facts, dates, and a clear understanding of the educational system as it pertained to minorities during the 1940s, 1950s, and 1960s.

He is a loyal family member who worries about his nine living siblings—eight brothers and one sister—as well as his mother. He has a special connectedness to the land, the place, and the people, especially the Pima Indian heritage of his mother. He is also a member of the Tribal Council on the nearby Gila Reservation. His voice is filled with a vision of what he needed growing up but did not get, of promises denied, the rights and wrongs about America, the treatment of minorities and his people—black and Indian. He spent a good deal of time relating how the educational system failed him and others of his generation in Randolph and the surrounding communities. He feels that he'd be in a different class and situation if prejudices had not interfered with his early years. Even though he is one of a handful of Casa Grande youths that went to college, he insists that his opportunities were limited because of racism.

He says: "Comin' up . . . I'd see things and think, 'What is this? . . . this isn't right.'" At age fifty-one, Ron echoes some of the same commitment of a Jesse Jackson, a Cesar Chavez, a Russell Means, other revolutionaries, and his brother Jeff, who see the bigger picture and the fallacies within the system. Like the others, he also retains hope, optimism, and a sense of humor to ride out the bumps and bruises of life.

Ron, interestingly, also wants to create a museum in Randolph to house the artifacts of the lives of those workers and migrant families who, like his grandfather Jesse Jordan and his father, Cleveland Jordan, came West seeking a better life and opportunities. It would be his way of helping to document their contribution, not only to the town and the region, but also to the economy and culture of the Casa Grande Valley and the American West. He is the philosopher from the cotton fields.

Our family is from Arkansas; I'd like to go back. We have relatives still back there. Hope and St. John, and Nashville, Arkansas. They are all in that corner near Texarkana. We have a lot of relatives there. As a matter of fact, one family came through, and they stopped at Mama's and she wasn't there and they talked with Jeff and they said, "We're your cousins."

My dad shopped at Randolph for the cotton workers, especially when they lived at Eleven Mile Corner. You didn't go to the theater that much, so for recreation my dad had a baseball team. It was mainly made up of the workers. They'd level out a spot up there, and they'd call it their diamond. After he got tied up with Mama and met people on the reservation and other camps, then they had a little league.

He used to have his own uniforms, bats, balls, and all the people would get in trucks and we'd travel to Blackwater or Sacaton, and they'd play each other on the weekends, on Saturday afternoon or Sundays. When we moved to Randolph, they just kinda quit.

There was another family, the Hendersons, at Eleven Mile Corner. It's a big rock house with all the trees around it. There used to be a big camp there. Mr. Henderson had a team too, and we used to play them. It was always rivalry to play them because we were all real good friends. They played ball and sold pops and food stuff, that was the way they had a good time. Other than people drinking beer and dancing, there wasn't much. Every camp that my father had, he always had a little place where they sold hamburgers and pops.

On Saturday night, they'd be playin' poker or smokin' in the cabins. In front of the big cabin, Miss Lola, I still remember her, would make you hamburgers, or they'd have chips and stuff. I think they had a bar or two then. If you didn't have money, you had a charge account, so you could get a hamburger. It was twenty-five cents, so when the next payday come, they just deduct that. I remember sometime the guys would just live day-to-day. Toby and Tokay and Booker and all them. He was named Tokay because he drank Tokay wine. They work all week but ate and drank up a lot of what their bill would be. Almost more than what they made. Friday or Saturday after work, they'd buy 'em a bottle of white port or Tokay wine. That was their life. My father didn't sell alcohol. They bought that at La Palma or somewhere else.

Before goin' to Randolph, we lived at Eleven Mile Corner about a mile south and then a half mile west of Eleven Mile Corner. The homes and the camps are not there anymore. That's when we started school. My sisters, Pearl and Bessie, and brothers Georgie and Willie and me all went to Eleven Mile Corner School. Mr. Batteau was the teacher. That place where they taught us was like two little separate rooms, kind of like a little auditorium. That facility was really built for the workers in the camp. All of that was federal or government, somehow.

The younger kids, like kindergarten, first grade, second grade, didn't have a teacher. Mrs. Batteau volunteered just on her own time and taught us. So we were real close to them, and thankful for them, for how they got all of us into school, tryin' to educate us. After that, daddy bought that place in Randolph and we moved over there I think in 1950 or '51. From that point, we no longer went to Eleven Mile Corner School.

But, the Eleven Mile Corner School was built for minorities. There was no white children there. As far as I remember, it was just all black. But the kids that did go to school there lived there in that camp facility, and some, like us, came from the outside, fringe areas. We didn't have no bus to pick us up. We walked about a mile and a half, two miles, back in them times. When we left there and went to Randolph, to that new house, the Coolidge school system buses picked us up, and we went to Borree Corner School. I still remember

Mrs. Neal, she was the principal, the teacher, and everything. And Mrs. Barnes. They were just like your parents.

Borree was all segregated, too. I have never seen so many black kids! WOW! Where did all these guys come from! [Laughter] I was still a spot in the crowd. You know, little, straight hair, light-complexion kid, and there's all those black kids. But I knew that everybody knew me. I was never mistreated. I was called "cat eyes" a lot of times, because of my green eyes. But I was never mistreated. Actually, I was closer to the blacks than, say, my Indian relatives. They were a little distant, you know. We were very intelligent kids, because we were forced to be intelligent kids.

We had spelling, arithmetic, geography, or history. And every day we had homework. Every day. And we'd go home and do our homework, and if we didn't have it or if we missed on the test, we got a whippin'—a lick for everything we missed.

All the teachers at Borree Corner were black. Besides Mrs. Neal and Mrs. Barnes, there was two ladies named Mrs. Dixon. Big Mrs. Dixon and little Mrs. Dixon. I don't even know where they lived or came from, and we never knew no first names. You just never used a first name for a teacher or an elder. The light-complected teacher, Mrs. Barnes, was pretty. As a matter of fact, Mrs. Barnes and little Mrs. Dixon, they were real pretty. Well, I would see all of the eighth-grade boys, you know, all lookin' at 'em. [Laughter]

But, them teachers was FIRM. They was just strictly business, and I think they had to be strict, no playin' around. There was a whipping every day. Matter of fact, recess is when you got your whippin'. Me and this other little girl, Liz M. Banks, and certain ones were always in trouble. We knew we was gonna get our whippin' either from misspellin' a word or two, a math problem, or somethin'. And they'd tell you what you did and that's how many licks you got. And they were hard licks! You had welts on your back when you went home, and nobody laughed at you—nobody.

The rest of the kids would go on and play. When you get through whimperin' then you go on and play with 'em, because they know that the next day or so, they was gonna be in there too. The Miles family lived at Desert Beach. Joyce Miles, Boodie, Claudel, and all them, they were real smart. Joyce was kinda my friend. And once they did their work, they tried to help you to prevent you from gettin' a whippin'. Sometimes, at recess or if we just by chance happen to crisscross through workin' or somethin', they would try to help you. I never forgot that. Even today, she lives in Los Angeles now, if I call Joyce now, she'll appreciate it. I'd say, "You saved me a bunch of whippin's." We'd laugh about it now. We were, I think durin' them times, in second, third, and fourth grade.

We were smarter black children because Mrs. Neal and all the others made you learn. When I went to the fifth grade was when they segregated the schools. Then they closed Borree Corner School, and we all went to Kenilworth School. I don't know where some of the other kids went. I know some went to Coolidge, and I think some went to Eloy. Actually, some of those kids came all the way from the Eloy and Florence area, from all them workin' farms. They came a long ways.

I didn't know anything about Eloy when I was growing up. I was still a kid. All I remember about it was later, goin' over there, and I didn't like it because everything looked so bad. Eloy was a real rough town. People got killed every weekend. There's even books on that. Eloy, Arizona, how it used to be! Drinkin' and havin' a good time! It was just work and live day-to-day. You could whip somebody, and shoot 'em, that was just part of it. But when they segregated the schools and we started goin' to Kenilworth, there was a lot of problems. The whites called you names, and you ended up callin' them names, and the next thing you know, you're out there fightin'. And the teachers, they'd whip you. I think that, sometimes, they would whip you more because they could just do it.

And even the bus driver, "Old man Staggs" was his name. He'd pick the kids up in Randolph, and he was a hateful, prejudiced man. And you better sit down and act right on that bus. They treated you like slaves. And when you rode from Randolph to go to school, you didn't take the Coolidge route, we took all back roads. And we'd be wantin' to go through Coolidge to see the town, you know. Sometimes the kids would be actin' up, so he'd just pick somebody out and pull 'em off his bus and pull his belt off and just whip 'em real hard. We didn't like that, but you was scared to do anything. Many times a lot of us thought about just jumpin' on him. Even when he'd park at school, if you'd get off the bus too slow, he would have his belt off and he'd be hittin' you. I remember that real well.

He was the caretaker of the school at Kenilworth. I think that's when the downfall of a lot of us start hittin', because most of the teachers were white, and most all of us sat in the back. You didn't get the attention that you did at Borree Corner School under Mrs. Neal and Barnes and Mr. Batteau. All the kids that graduated out of Borree Corner School out of the eighth grade went to the white high school in Coolidge. They did pretty well in that school. But when we went through we didn't get the attention as much. I think it was for a couple of years or so. Then it started mellowing out, and everybody started to be treated a little better. When I left Kenilworth and saw how things was, I didn't like it. So I went back to Coolidge because, in Coolidge, they didn't know if I was black or what, so I was treated like the others.

The track team from the integrated Kenilworth School, 1960–1961.

Some of the teachers they brought in were newer teachers, and one in particular was Dr. Sams. He was a new teacher, and that's when I started runnin' track. Even after I left his class and went to the next grade he would always come around and look after me and see if I was okay. So I had good teaching, through the school system. The only bad thing was when I went to play ball.

My eighth-grade year, I went back to Kenilworth, 'cause things were a little bit better then. The eighth-grade class only had ten of us. They had a good teacher there, Mr. Solomon. He was a good, fair teacher, so I got a little better education out there than I would have had in my eighth-grade year in Coolidge. Actually, I kind of played my hand between the systems and schools to see what's best for Ronnie. And so, when I finished there I came back to the high school.

I remember desegregation in the 1950s. We didn't know what to expect. They had a little candy store where you could buy candies and other things, and the white kids had some little licorice babies. They called it "nigger babies." They'd say, "Give me ten of them nigger babies." They'd be doin' it just to aggravate you, you know. Oh, there was some fights. That was the first time that we got associated with the Mexicans and Indians because they were also put in the schools. From Randolph, blacks had been goin' to Borree Corner School; some Mexican families were livin' here too. And so when they did that, all of this section went to Kenilworth, that included the Mexicans and everybody. My brother Berdette, and the others younger than me, didn't go through as much as I did. I don't think so. That's why sometime it's hard being a mixed or integrated kid in those times.

As far as education goes, when the black girls finish high school, they were just like the boys, but most did not get further because of startin' families and such. When I was in school, I took math. I liked math because of my dad's business. We have to add columns of figures for the people who worked with us, whether you picked cotton or chopped. We had ledgers. And I used to have to add them in my head because we didn't have machines or nothin'. And I took algebra and got pretty good at it. And then my junior year and my senior year, I wanted to try trigonometry. When I went into the classrooms, seems like it was okay for you to take first-level algebra, but certain other classes, like trig, or geometry, higher-up classes, college-prep classes, I got the feelin' the teachers that taught those classes didn't want you to be there. I'm sure that even some of the kids in our school system, like blacks, just knew not to take that class, 'cause they didn't want you in there to start with. If you took it, you gonna fail or they didn't help you.

I'll never forget this old lady, I called her old lady Conrad. Her name was Mrs. Conrad. She was a very hateful lady, but I didn't know that she was that hateful until she was a teacher in that trigonometry class. I had a good friend from playin' ball in the high school system, his name is Jerry Jones, he was real good to me. And Anita Jo, she was in that class. And they knew that the teacher wasn't gonna help you or do things for you. What they'd do is, when I'm in class and I'd raise my hand and want to ask a question, she wouldn't acknowledge you. And it hurt, you know. I'd sit there and the kids would look at you. And they knew. And they would just kinda bow their heads. When we had study hall, it was mandatory. They would get to the table and help me. And I passed. She was a hateful old woman, I'll never forget that. But I didn't know that she would be hateful and prejudiced that way.

I had Algebra I and II, and wanted to go to the next level. Same way with chemistry. Our chemistry instructor, Mr. Ronarud. The first part of it was okay. And then as time went on I wanted to get to the higher class. Then I noticed things were kind of changing. I was the only minority in there. There would be a Mexican or two, but never blacks. And so comin' up, the school system held us back. I joined the Indian Club, they called it the Amerind Club, for a couple of years. The lady that was the sponsor of that club gave you information. They have so-called Indian counselors about furthering your education. She would never tell me nothin'.

Later when I came back and I asked the teachers, "How come that lady never told us about what was available," they wouldn't answer or tell you because they knew that they didn't want Indians gettin' better. And of course, me bein' mixed with black, they didn't want you to know what was available, so you couldn't go on to the university. That kept me down, but I had to leave this state, to go to California 'cause things were a little different there. There was a few spots that was a little rough, but it was better. I got better treatment out there and everything, and doors were open to you. I'm still bitter, though, about certain things around here.

So a teacher like Mr. Batteau gave us the chance to get started in school. And as I learned later, he was more than just a teacher, he was kinda like a father image to us too, all of us, not only our family, but other families. They all looked up to Mr. Batteau and Mrs. Batteau as somebody that really cared for us, and wanted to educate us. They went out of their way to do things for us.

Most all the teams that played out here were predominantly black. Because you had all the Randolph kids goin' to school over there at Kenilworth. So we come back and these kids knew me then. And we'd bring our Kenilworth team over here and our coach was Mr. Raul García. He was just a new teacher at the time, and we'd beat 'em. I laughed at 'em, because even the coach, you know, says I wasn't good enough to play on that team, but yet we played here and we come and beat 'em. And we'd beat Casa Grande School, too. We actually were Randolph kids. Randolph was actually beatin' 'em! You know, little black kids—they didn't like that.

When we first went to Coolidge High School, we didn't really know anything about tackle football. I had never been to a high school game where they played tackle football and wore helmets and shoulder pads and all that stuff. So when we came from Kenilworth in our freshman year, they said, "You gonna play football?" And we didn't know any better; we said, "Yeah we gonna play." And here I was a little guy, and goin' up there and we didn't know names of positions like guard, tackle, ends, and center, running backs or quarterbacks, or anything. Of course, we played an eight-man team, flag football, but it was nothin' like up there. So we're all sittin' outside talkin', well . . . "What spot you gonna play?" I said, "I don't know, I'll get on the line I guess." So the coaches come out and they tell us and we're talkin' and said, well, "You oughta be the runnin' back. You oughta run with the ball." I said, "Well you get hit all the time." I said, "I don't wanna do that, you get hurt! I'd just rather be on the line and knock the guy out of the way." [Laughter] And so the coaches came out and as time went along you learned; I ended up bein' a running back.

I had a good coach from Arkansas, Mr. Keesler. He was my second coach. The first-year coach, he came out of the University of Oklahoma. But he didn't seem to show any concern really for the team or for the players. He kind of landed a job and wanted to move on. But after his first year, Mr. Keesler came in and he was from Texarkana, Arkansas, and when he learned that our family was from Hope, Arkansas, he showed a little interest in helpin' me, and I guess he just identified with my dad. He didn't really talk to him a lot, he'd just come out and ask him if I'd go to practice. I guess 'cause I was a pretty good athlete. As far as clothing and stuff goes, my dad didn't have money for that, so he had to go to the Coolidge Cleaners and talk to the people, and told them my situation. Mr. Tias was the owner then. So he sent word through school to stop by and see him, and I'd go by the cleaners and he'd say, "Go back there, and whatever you can wear and whatever will fit you, you take it with you." And that was all because of Mr. Keesler. He also invited me to his home. And, before Friday-night games, instead of me comin' all the way to Randolph, and then have to hustle a ride back to play ball that night, he'd take me to his home. And they'd feed me a big steak dinner. I'll never forget that.

Track season, I had another good coach, Mr. Hawkins, who is still in Coolidge today. He basically did the same thing. He kind of took care of me too. The only thing that I look back on now, and that I didn't receive then, was education, and I told him about it. When I first moved back from California he came by and looked me up. I told him that as I left Coolidge School and went out to California and learned a little more about the educational system I asked, "Why, when I was in high school, why didn't you try to help me or guide me to the University of Arizona or Arizona State or different places, whereas the white guys got all the instructions on what to do and how to do it." They never mistreated me or nothin' like that, but at the same time, as far as I can remember, nobody went out of their way to help the better black athletes or tell you about what to do or how to go about doin' things. That never, never occurred.

Another good black athlete, L. B. Reece, went to Kenilworth Grade School, but he never attended high school. Fannie Mae and his dad worked in the fields a lot, so he didn't really have the home life which encourages a kid to go on to school. But L. B. was a great athlete. That man could have probably gone on if he wanted to get good training. He could have gone to the Olympics, I think. After eighth grade and you graduated from junior high school, parents thought that that was enough. You can count money, so let's get on this field and go to work. And so they figured you could go and work in the fields rather than further your education.

I was the first one of the family to go to high school. Well, Bennie went too. Me and Bennie, we graduated from the eighth grade in Kenilworth School, and then we went the freshman year and Bennie went up to his sophomore year, then he quit. I guess he kinda saw that things were a little segregated more at that time, and he was maturing and learning how things were. Then his mother, Nona, the first wife of my father, Cleveland, moved to Los Angeles. And so, somehow or another Bennie ended up goin' out to Los Angeles with his family there. He saw things better there. Eventually, he made a couple trips back and forth, but I had never gone. I was pretty popular in high school, and I wanted to finish it here. But when I finished school, Bennie told me about it, and then our other older brother, Woostie, had been goin' to Riverside, California, workin' potatoes. My junior year my Aunt Billie came out and took us, me and Bennie, to Los Angeles. That's when I first saw that big pretty city. I say, "I know it's got to be better here than over there." And so when I finished school my dad didn't really approve of any of us goin' out there. He always thought the city was a bad place because you never will make nothin' out of yourselves there. I heard him say that, so I was determined to make it. Because when I finished here, there was no college or University of Arizona or Arizona State interested in me. And Eastern Arizona College at Thatcher was there, but it was like a small place. It's a community college, and a lot of whites went there. The only Indians up there was Apaches, and I don't think too many went there 'cause they either went to Northern Arizona University or Phoenix College. Some went to that Indian college in Kansas called Haskell. So all the Indians had their own routes to follow.

I could have gone to Sherman, but nobody ever told us about it. They had it set up where you could go and learn trades instead of gettin' into education or other fields. But we never knew that. I didn't even know all this was available to me until I was a full-fledged adult. Then I only learned through conversations with other people that went that route. When I asked, they said, "Well, didn't you know that?" I said, "Nobody ever told us." All in my family qualified as being Indians, because we were more than one-quarter. We are three-eighths. But we didn't know what benefits could come from that.

After my younger days, and I left my grandparents and went through grade school, I'd go out there and stay with 'em and come back as I got older. It was kinda dead on the reservation, so I started socializing and traveling a lot more. I never went back, and we never knew what was available to us. Our family was stuck way out here in Randolph, and we did not associate so much with the Indians. I did, but not to the extent of what was available to me. The only somebody that gained from that was my sister Farietta in San Diego, because she was sent to San Diego Mesa College. It was her choice to go there. They paid for all of her expenses—apartment, spending money, everything. All she had to do was just attend school.

I had all that available to me but didn't know it. I was determined to go to Compton College because of this one guy that went to University of Arizona who came through the Coolidge school system and kinda explained how he did it. Bobby Lee Thompson was his name. He was a football player at the University of Arizona. He was a black guy. Anybody that went out to California could go to that community college without paying out-of-state fees. You just paid tuition and you bought your books. So I went out there and tried for the football team and made the team. But because I was out-of-state, I had to wait a year. For one year I foot the bill for myself. The only people that showed concerns was the white farmer that lived across the tracks from my mom, Rodney Kleck. Kleck Road and Kleck Farms is named after him. Now it's a big refinery over here somewhere on this side. Mr. Kleck used to bring his kids up to where that dirt road is, to turn to Mama's, and drop his kids off to catch the bus to go to Coolidge School. When I was goin' to Coolidge School, he'd always ask me, "Ronnie, watch my kids, and make sure they get on the bus and straighten up." And so I watched his children for him. When I finished high school, I worked for him too; we took care of his fields. So when I left and he asked Dad where was I and where was I goin' to school, they told him I was in California tryin' to go to school. He would then tell me to let him know how much money I needed. But I felt bad takin' Mr. Kleck's money, you know, 'cause I was stayin' with Bessie, and I paid her whatever I could pay and I bought my own books. But he still sent me money every month. When I got to my sophomore year, I told him, "Mr. Kleck, that's okay, I'm fine, I don't want to take your money, I'm doin' okay." You see, I hustled on my own, and I felt bad, I didn't want to take somebody's money.

Mr. Kleck farmed cotton, even today. All cotton. We chopped his cotton and picked it. My dad worked for certain farmers, like Kleck Farms, Rodney Kleck, Mr. Ladd, Austin Ladd farms, and others. But there was certain farms that we took care of every year. Our dad had his own contracting business. We also irrigated their fields too, but each farmer, like Mr. Kleck, was the big boss, and had a foreman that actually did the work. He was the guy to sign the check and did everything. But the foreman was the guy that oversaw you and told you what to do. The worker never dealt with the farmer directly. The only somebody that dealt with the farmer directly would be my dad. He would go and talk with him or pick up a check, and a lot of times, he would come by the field once in a while or maybe drop a check off, but normally that never occurred.

The foreman was the guy that actually managed the farm. He was the one that hired the irrigators, cultivated the fields, had people drive tractors, cultivate, and do all that kind of work. I think it was some tension there, because every time you saw the owner or the main farmer come up, everybody got busy or got nervous. And you didn't stay around the truck to get water too long, you was always gettin' out of the way. So some tension was there because it was the fear of the big white guy, the big owner. It was like that even when you went to the Texaco station or the Woods place or the big people up there, you always had that little bit of fear, you know. You just didn't go in the station and, like they do now, self-serve and get your gas and pay your money and stuff. You got permission. "Mr. . . ." I can't remember what we called him " . . . can we get five dollars of gas?" or more like two. You never bought five dollars of gas 'cause gas was only 20 or 30 cents. "Can we get two dollars of gas?" So he'd say, "Go ahead." And then it was always, "Thank you sir." "Yes sir." Stuff like that. I remember that if the Texaco didn't have something, you went over here to the Chevron station and old man Thompson, who was a mean guy. In that real nice bar where they sold whiskey and had a nice bar and stuff, they sold eggs, sausages in the vinegar solution, and all that stuff. I remember seeing people that wouldn't go in there. They would wait for him to come out to see what you wanted. And then you asked and then he would tell you no, or he might be able to help you. If he didn't like you, he'd tell you no. You just kinda bowed your head and just kinda walked away. You never said nothing. He was very prejudiced.

Whereas the Marshalls were a little more friendlier. But see, in those times you had to realize that whites had a certain attitude. They had to keep among themselves. As far as treating us, they kept that certain distance from the blacks, and I think in some cases they treated the Mexicans a little better. The Quinteros and certain ones could go in here and do things, whereas we couldn't. Comin' up, and bein' a kid, I'd run with these kids and I'd see things and think, "What is this? . . . this isn't right." And I still see stuff today that I say is not right. But you can't change it, you know. All in all, we worked in the fields, and when the farmer came around, it was that fear. But the farmer never really got on you directly or nothin' like that. And sometimes we would be workin' out there and Daddy would pull up with all of us in the truck or in the bus. He didn't pull too close to his house. When you pulled in there, you pulled out of the way, you just didn't get off and go sit under the guy's tree. You knew to keep that distance.

So in between all that workin', I was goin' to college at Compton. Well, when I went to my second year there and I kinda got to know California, I started finding other people that I knew from here, from Eloy, like the Case family, lot of them was from Eloy. After school in Casa Grande or Coolidge, Eloy or Florence, blacks had nothin' else to do, unless you worked in the fields. The late 1950s goin' into the '60s you could see the farmwork goin' out because of automation of the new cotton pickers. Somebody invented that cotton picker that go and pick that cotton off those stalks. So there was no need for the human being to pick it no more. So all us younger people, you had to do somethin'. So when Bessie called up and said that I could go to school in Los Angeles and stay with them, I kinda learned what's what. I made it there for three semesters. That was the same time the Vietnamese War started. And so over half of our football team that had come to college got drafted. We all got drafted out of school. That was in 1966. We were right in the middle of the fall semester, and I didn't even attend the spring semester; I had about a week or so left, so we all just reported for service. I got out in May of 1969.

After that I went back to Compton College and took a couple of night classes, 'cause I had gotten married. Then I bought a home in Carson, California, which is south of Compton, California. And that district was Los Angeles Harbor College. At Los Angeles Harbor College, I got an "A" in business administration. That same time is when I joined that company that I worked for when I was at Compton College. Big Donuts. When they went into the Pup 'n Taco business, they called me up and put me back to workin'. I worked with them from day one until 1986, when they sold out. But I was makin' big money. So you can say I started with them in June of 1969 through 1986. In the middle 1970s, I was a manager, then later as a supervisor I would probably gross let's say about $25,000 a year, and then it gradually went into the thirties, and then when I get to the late 1970s up into the 1980s I was grossin' over fifty-some thousand a year. I was makin' good money. Of course I worked hard and had a lot of responsibility. Durin' that time, they kinda frowned on you wantin' to go back to school 'cause I was makin' probably more money than a lot of professors or other people.

My children and my wife, then, were livin' a good life. I was livin' a good life too. And I sent monies home to Mom here, or I'd take her on trips to San Francisco and visit family up there. Or just take her on vacation. When that company sold, and I had always wanted to come back here to Arizona and be around Mom because she's gettin' up in age, I also wanted to get to know my younger brothers. I really didn't know them that well, 'cause I had been away from them. The youngest brother, Adrian, I didn't know him at all. I'm actually closer to Bessie and Bennie in Los Angeles than here, you know. But I think as time goes on, we're gettin' closer. I think that I had drive in me when I went out there. From listenin' to my dad sayin' nobody will ever make it, and I was determined to make it. See my goal was, I was gonna go there two years and then transfer to the University of Arizona. Like Bobbie Lee Thompson. But I never made it 'cause I got drafted into the service and had to go to war. So, that ended that, but I did go back and get the associate's degree (A.A.) in business administration

I never imagined that it would be like this, that I'd come back here and to Arizona. When you make a résumé and you state "Director of Operations," you're makin' all these big monies, and livin' high off the hog and stuff like that, you can feel the resentment, right today. People wouldn't talk to me. They wouldn't. I went to Coca-Cola, I thought I'd have a shot with them because I knew one of their big people out of Atlanta, Georgia, and I put his name on the résumé as a reference. Those people, they wouldn't even talk to me. Pepsi-Cola, a lot of the bigger companies here, nobody would talk to me. And so finally somebody says, "Well, do you know why, because you were a director of a company. You were makin' big money, and you were a minority. You just not gonna make it here."

So the only way I got back here and found suitable employment was through the Bureau of Indian Affairs, because of my Indian blood. And I took a ground-level job, as an electrician's helper, because I knew a little basic stuff, but not enough to go out and get an electrician's job. I also knew that in order to get established around here, that would be my only chance. Start from the bottom. So now I'm, well, I bought this house.

Overall, in this county, the black man still has difficulty makin' it. Only areas that I've seen them have opportunities was when they said that you had to have a certain amount of minorities workin'. In the highway department, I think they got one or two blacks workin'. I've only seen one, his name is Isaac. I don't see no others on the highway department workin'. The sheriff's department, Melvin Moore, he's been with them for years. And so I think that comes about because they need a black or somebody to kinda go and look after the black community. Around here, at the electric utilities district number two, there's no blacks employed at all. They're all white.

I think maybe my brother would have a chance to work there if our company divested. But the guy that's headin' that up now, I just asked him point-blank, "Are they prejudiced out there?" And he says, "Yeah. That man said that he'd never hire a black." So it don't do you any good to even put in an application. Out there, you can't get a position. Southwest Gas, where Moochie works now,

I think they have one black guy there, and he's in maintenance. But the only way Moochie got on with that company is that he had worked with the Indian Tribal Council and the gas company took over the tribe, so the tribe asked if their people could get a chance to work. And so, that's how Moochie got hired. I've never seen a black guy in the Arizona Public Service, around here. I've seen one black guy that was workin' in Coolidge, but I think they came from Phoenix or somewhere to work. Then they've laid off so many thousands of their people. So if a minority goes there to work, you're not gonna get on. I applied for Southwest Gas, but they're not a strong believer in nepotism, being that Moochie worked there. They just tell you, "Well, you got one here now. There ain't gonna be not one more of ya."

A young black person, unless you just stick with that school and go through it, you just got obstacles. It's rough. My daughter is at the U of A in media arts. I actually told her to finish there, but she wanted to go to ASU. I told her, I says, "You'll be better off just to go back East to the University of Maryland, or somewhere else, where minorities have a better chance than here." But she really don't want to go back there. Her aunt is back there. She's been tryin' to encourage her to come back there. She's afraid of the cold and all that too. [Laughter]

My mom had trouble when my father proposed to her, that's what I heard. Well, the tribes and the Indian people were also regulated by federal laws. Actually, the Indian was kind of owned by the government. They were just like, treated like animals. You go out here and you just look back on things like I do now, at some of these reservations, you'll feel something in your stomach. Me and Marie went up to Utah to Zion National Park, and I wanted to go through the Navajo Reservation because I had never seen it. I looked across that barren desert and there's nothin' out there but mesas. They told the people, this is your homelands. How in the world they gonna live? They just stuck them out there and didn't care nothin' about 'em. But the only advantage for those Navajos is, I guess, the minerals. There's a lot of coal and uranium up there, so now they're pretty wealthy. So through political ways and through laws bein' changed, they gettin' back at them. They're taking their monies. Whereas our reservation here used to be beautiful because the Gila River ran through it. They irrigated a lot. They farmed a lot. They built that Coolidge Dam and they dried up all this area. So much of those waters was supposed to go to the tribes, and so much to the white man, but most all of it went to the white man. The Indians got some, sure, but it never was a fair balance, and it's still not that way today. Oh, it's just unfair how things happen, how they treatin' the people.

Epilogue

It has been ten years since I first rode up to Randolph filled with anxiety and expectation about what I would find there. As I "conclude" or attempt to bring some closure to this project on Black migrant life in the Casa Grande Valley, I find myself working very slowly on the final portions of it. It's as if I am reluctant to leave all those faces, voices, smiles, laughs, tears, yards, houses, empty fields, and empty roads that comprise all those communities—Casa Grande, Eloy, Coolidge, Eleven Mile Corner Road, Randolph. The rubble of the Borree Corner one-room segregated school is barely discernable. A tired-looking old cotton gin stands on the opposite corner facing west, majestically.

No matter how many times I have driven those sixty-five miles from Tucson, I always become overwhelmed with emotion. I never drive fast along Routes 84, 87, 287, 60, or the Selma Highway, but travel slowly, to savor and experience every minute of every image my mind creates. Many of the cotton farms now have homes on them or are not planted at all. Others still have acres and acres of thriving cotton from March to October—from planting to harvesting. Instead of hundreds of black, brown, yellow, or white bodies bending, talking, picking, singing, or sweating, the picking machines powerfully dominate the landscape. I dismiss them and put in their places the visages of the people, the working people, my people. I am not a romantic, yet I feel so attached to this land and these families that I often ask myself, Why is it that I, a Caribbean woman from peasant and red-dirt roots, would come to this place to document and connect so strongly with a portion of America's dispossessed? My daughter, whom I've always said was a born "social worker," has often said this: "Mom, if you can't make a difference or give back in this world, what's the point"!

I always knew that the elders of these communities would be my best and most truthful resources, so the urgency of getting their stories was always a priority. That proved to be the case as their various illnesses interfered with their memories and bodies. I never got to meet Clara Bell, Randolph's midwife for whom a street is named; when she became ill her children took her to California to live with them, and she subsequently died there. Mrs. Mathis, another midwife, also passed. Edna Fields and Lucille Smith died while I was out of the country. Myrtle Jordan, after a lingering illness and lung disease perhaps hastened by the various chemical and toxic pollutants brought into the community, also died. I had thought she'd live forever because she was so spunky and invincible! Christine Moore Fields, who carried the legacy of two matriarchs, Beulah and Edna, died; and Mamie Selmon, the oldest Sojourner with her unforgettable eyes and tenacity, also passed—both in 1999. Mr. Taylor and Mr. Muldrew passed; Pearl Jordan (McKinney) passed valiantly holding on to her true Arkansan spirit. G. L. Turner moved away when he could no longer keep up his properties and take care of his ailing wife. And O. C. Fields and Fannie Reece are ailing.

Randolph has been dying as well, along with its elders. There seem to be fewer houses now than when I first visited in 1989. The yards look lonelier, and the children's voices are more muted, despite a playground built on an eastside lot. The "new" trailers that have appeared, seemingly overnight, house new Mexican residents who add little to the life of this shrinking town. The Moore convenience store closed in 1995—a sign of the town's further decline. Some residents still work at the Arizona Training Center across the road, but it, too, has its limitations; some residents of working age leave to look for other opportunities or just sit. The Sunbelt Refinery closed due to statewide public pressure—or had it become economically unfeasible for the powerful? The Proler Tin Manufacturing Company is still there and recently contracted to become a recycling plant for hazardous wastes from several states and Mexico. That is very sad, because already at least four deaths have been attributed to the presence of toxic fumes from various polluting industries in and around Randolph. The "dumping on the poor" continues; the boarded-up stores along Highway 87 are still there, sitting lifelessly. The Roof Machine Shop continues to barely operate. Whites still live on the west side of the highway and the colored ones on the east side; and even now, in 1999, one wonders why.

There is little optimism in these towns, so life is lived one day at a time. Some, like Cleveland and Jesse Jordan, Beulah Moore of Randolph, Georgia Rupe of Eloy, and the Fieldses of Coolidge, did go from "sacks to riches"; but others, such as Fannie Reece, Nick White, and Jeff Jordan, think that Randolph, like some other valley communities—Vicksburg, Mobile, Cashion—"ain't got long!" Despite these conclusions, they have all left their imprint on the Western lands, economy, and culture. And, like the Exodusters, Okies, and Buffalo soldiers, they will become the stuff of legends, leaving the legacy of their determination, hard work, and strong spirit in their children.

The grandsons of black Okies in Randolph today.

Notes

Prologue

1. Jonathan Hicks, "Built on Cotton, Town Feels More Threadbare Daily."

2. Geta LeSeur, "Dreams Deferred: The Women of Randolph, Arizona."

3. These numbers are not meant to be absolute; rather, this is the approximation I heard from the residents. The county seat in Florence, however, shows completely different figures because its numbers are based on voter rolls. Those records do not include children.

4. John Thompson and Cary Wolinsky, "Cotton, King of Fibers."

I. "There Are Pioneer Souls That Blaze a Path": Dreamin' the West

1. Susan Wloszczyna, "Film Aims to Widen Range of Westerns."

2. Charlotte Lowe, "True West: Black Cowboy's Pioneer Spirit Has Built Ranch Town of Keeylocko."

3. Wloszczyna, "Film Aims."

4. "First Families: Early Black Pioneers."

5. "First Negros in the Salt River Valley."

6. Richard E. Harris, *The First 100 Years: A History of Arizona Blacks,* 4–7.

7. Ibid.

8. The term and classification *Indian* are used in this study to refer to several tribal groups in Arizona: the Pima, Papago, Navajo, Zuni, and Yaqui. Those who picked cotton or lived in the Casa Grande Valley were primarily Papagos and Pimas. In the narratives, and in speaking, everyone always said "Indian." For example, "The Indians picked with us but kept to themselves." *Indian* was also the group descriptor of choice before the 1970s, as was *Mexican,* which at the present time has been replaced by *Hispanic. Mexican* and *Indian,* in Arizona, remain the common usages by all groups.

9. William Loren Katz, *The Black West,* 170.

10. Ibid.

11. Thomas C. Cox, *Blacks in Topeka, Kansas, 1865–1915: A Social History,* 38, 40.

12. I have coined the phrase *black Okies* to differentiate black migrants from white migrants. They came primarily from the states of Oklahoma, Kansas, Arkansas, and Texas on similar journeys for work and to escape the devastated crops and land in the 1930s. White Okies, however, are classed as those white migrants who came primarily from Oklahoma. Black Okies were *not* landowners but workers on white farms and cotton fields.

13. *Colorado Prospector* 20:2 (March 1989): 1.

II. "Men That Press On with the Ardor of Hope"

1. Harris, *First 100 Years,* 40.

2. Ibid.

3. Abraham S. Chanin, "McNary—a Transplanted Town."

4. Ibid.

5. "The Romance of Randolph, *Arizona Republic,* 1925, anonymous advertisement, Florence Historical Society.

6. Will C. Barnes, *Arizona Place Names.*

7. This information was gleaned from Sarah Lucille Smith's personal files. She had access to such information as a protégé of the town's first postmaster, Channing E. Babbitt, and then as the Randolph postmaster herself.

8. Information in the following paragraphs is from Sarah Lucille Smith's personal file.

9. Katherine Hupfel, "Main Street Medley."

10. Hupfel, "Randolph and La Palma."

11. David Hoye, "Dumping on the Poor," 1.

III. "Next to You, We Love Cotton Best!"

1.The Okies in *The Grapes of Wrath* were poor whites in search of new land and work. They traveled primarily in trucks along Route 66. Most of them ended up working on the fruit and vegetable farms in California. The novel widened the debate about what to do about the migrant poor.

2. Dorothea Lange is known internationally for her vivid and astonishing photographs of people. In the 1930s and 1940s, she went to the Casa Grande Valley in Arizona and did a series depicting the plight of migrants. The scenes of their living conditions sent shock waves to Washington, D.C., helping to initiate the construction of clean and habitable government camps and the creation of migrant opportunity programs. The Lange photographs are in the National Archives in Washington, D.C. See also Betsy Fahlman, *Cotton Culture: Dorothea Lange in Arizona* (forthcoming).

3. Obie Blackmon, "Sharecroppers in Southern Missouri."

4. Ibid.

5. Excerpt from Ken Light's *Delta Time: Mississippi Photographs* is quoted from "Witnessing the Lives of the Delta's People."

6. "Difficulties in Hiring Indians for Farm Labor Probed by Government Agencies in State."

7. Harlan Padfield and William E. Martin, *Farmers, Workers, and Machines: Technological and Social Change in Farm Industries in Arizona,* 88.

8. Ibid., 77.

9. Thompson and Wolinsky, "Cotton."

10. Ibid., 70–71.

11. Padfield and Martin, *Farmers, Workers, and Machines,* 84.

12. Excerpt taken from Mary Wilson's obituary, courtesy of Parlee Bradley. Wilson died on April 9, 1995, in Eloy.

13. Padfield and Martin, *Farmers, Workers, and Machines,* 102.

14. Dan Woods, "Cotton Pickers Wrote Chapter of Valley History."

15. Ibid, 2.

16. Ibid.

17. Ibid.

18. Ibid.

19. James E. Cook, "Cotton Picking Evolution Peaks in Sixties."

20. "Pickers Paid Millions."

IV. "We Are Sojourner's Daughters"

1. A longer version of this piece appeared as "Dreams Deferred: The Women of Randolph, Arizona," in *Compendium.*

2. Sharlene Hesse-Biber, "The Black Woman Worker: A Minority Group Perspective on Women at Work," 26.

3. Ibid., 27.

4. Sojourner Truth quoted in Elizabeth M. Almquist, *Minorities, Gender, and Work,* 61.

V. "Turn Not Away from Their Smiles and Their Tears"

1. Harris, *First 100 Years,* 59.

2. Quoted in John Yantis, "Casa Grande Looks Back at Segregation."

3. Irma Leos, interview, August 1997.

4. Matthew Batteau, "Child Accounting in the Eleven Mile Corner School." Batteau talked with me at length about his years of teaching migrant children at the Borree Corner School and the Eleven Mile Corner School, where he taught for a total of ten years. He also shared a copy of his thesis, from which much of my data comes.

5. Yantis, "Casa Grande Looks Back," 7.

6. Batteau, "Child Accounting," 2.

7. Batteau, "Child Accounting," 10.
8. Ibid., 12.
9. Ibid., 13.
10. Ibid., 39.
11. Harris, *First 100 Years,* 60.

Bibliography

Abajian, James. *Black Contributions to the American West.* Boston: G. K. Hall, 1974.

Agee, James. *Let Us Now Praise Famous Men.* Boston: Houghton Mifflin, 1941.

Allen, Ruth. *The Labor of Women in the Production of Cotton.* Austin: University of Texas Press, 1931.

Almquist, Elizabeth M. *Minorities, Gender, and Work.* Lexington, Mass.: D. C. Heath and Co., 1979.

Armitage, Susan Banfield, and Theresa and Jacobus Smart. "Black Women and Their Communities in Colorado." *Frontiers* 2:2.

Ashabramner, Brent. *Dark Harvest: Migrant Farm Workers in America.* New York: Dodd Mead and Co., 1985.

Barnes, Will C. *Arizona Place Names.* University of Arizona General Bulletin no. 2, 1935.

Batteau, Matthew. "Child Accounting in the Eleven Mile Corner School." Master's thesis, University of Arizona, 1935.

Benton, Pat. "Black Women Assert Role on Frontier." *Arizona Daily Star,* September 12, 1975, sec. C, p. 2.

Bernas-Elam, Jony. "School House Teaches ABCs of History: The Southside Colored Grammar School." *Casa Grande Dispatch,* June 1, 1992.

Bittle, William E., and Gilbert L. Geis. "Racial Fulfillment and the Rise of an All-Negro Community in Oklahoma." In *The Making of Black America,* ed. August Meier and Elliott M. Rudwick, vol. 1. New York: Atheneum, 1969.

Blackmon, Obie. "Sharecroppers in Southern Missouri." Paper, Black Studies Department, University of Missouri, 1995.

Bleiberg, Larry. "Almost Lost: History of Blacks in Oklahoma." *Columbus Dispatch,* February 14, 1999, sec. G, p. I-2.

Chanin, Abraham S. "McNary—a Transplanted Town." *Arizona Highways* 66:8 (August 1990): 30–35.

———. *This Land, These Voices: A Different View of Arizona History in the Words of Those Who Live It.* Flagstaff: Northland Press, 1977.

Coles, Robert. *Still Hungry in America.* New York: New American Library, 1968.

———. *Uprooted Children: The Early Life of Migrant Farm Workers.* Pittsburgh: University of Pittsburgh Press, 1970.

Collier, Randy. "Tiny Randolph Marks Christmas with Faith, Not Tinsel." *Arizona Republic,* December 14, 1993, sec. A, p. 8.

Collins, Karen Sikes. *Index to Arizoniana: The Journal of Arizona History, 1960–1964.* Tucson: Arizona Pioneers Historical Society, 1970.

"Colony of Blacks Invited to Homestead in Arizona." *Prescott Arizona Enterprise,* April 1878, pp. 2–4.

Cook, James E. "Cotton Picking Evolution Peaks in Sixties." *Arizona Republic,* October 19, 1986, sec. 1, p. 1.

"Cotton Chopping Gives Temporary Employment." *Coolidge Examiner,* April 30, 1948, sec. B, p. 6.

"Cotton Picking Machines to Be Tried in Eloy Area." *Coolidge Examiner,* July 23, 1948, p. 1.

Cox, Thomas C. *Blacks in Topeka, Kansas, 1865–1915: A Social History.* Baton Rouge: Louisiana State University Press, 1982.

Crockett, Norman L. *The Black Towns.* Lawrence: University Press of Kansas, 1979.

Crowe, John. *Discrimination, Poverty, and the Negro: Arizona in the National Context.* Tucson: University of Arizona Press, 1968.

"Difficulties of Hiring Indians for Farm Labor Probed by Government Agencies in State." *Coolidge Examiner,* June 1948.

Du Bois, W. E. B. *The Negro Common School.* Atlanta: Atlanta University Press, 1901.

"First Families: Early Black Pioneers." *Black Heritage* 1:5 (1976).

"First Negros in the Salt River Valley." *Sheriff* (December 1956): 52.

Grubbs, Donald H. *Cry from the Cotton: The Southern Tenant Farmers' Union and the New Deal.* Chapel Hill: University of North Carolina Press, 1971.

Harris, Richard E. *The First 100 Years: A History of Arizona Blacks.* Apache Junction, Ariz.: Relmo Press, 1983.

———. "The Homesteaders." *Black Heritage* 1:5, Arizona Bicentennial Issue (August 1976): 10–11.

Heltsley, Ernie. "More Industries, New Stores Foreseen for Oil Plant Site." *Arizona Daily Star,* October 2, 1989, sec. E, p. 9.

Hesse-Biber, Sharlene. "The Black Woman Worker: A Minority Group Perspective on Women at Work." *Sage* 3 (spring 1986): 26–33.

Hicks, Jonathan. "Built on Cotton, Town Feels More Threadbare Daily." *Arizona Daily Star,* November 16, 1980, sec. B, pp. 1–2.

Hill, Mozell C. "The All-Negro Communities of Oklahoma: The Natural History of a Social Movement." *Journal of Negro History* 31 (July 1946): 254–68.

Hoye, David. "Dumping on the Poor." *Phoenix Gazette,* April 25, 1993, sec. G, pp. 1–4.

Hronek, Pamela Clare, and Mary Logan Rothschild. *Doing What the Day Brought: An Oral History of Arizona Women.* Tucson: University of Arizona Press, 1992.

Hunts, Martin H. "Transiency and Its Relation to the Program of Pupils." Master's thesis, University of Arizona, 1936.

Hupfel, Katherine. "Main Street Medley." *Coolidge Examiner,* November 10, 1984, p. 3.

———. "Randolph and La Palma." *Coolidge Examiner,* December 29, 1950, n.p.

Jackson, T. C. "Negro Education in Arizona." Master's thesis, University of Arizona, 1941.

Jamerson, Elizabeth, and Susan Armitage, eds. *Writing the Range: Race, Class, and Culture in the Women's West.* Norman: University of Oklahoma Press, 1997.

Jones, Jacqueline. *Labor of Love, Labor of Sorrow: Black Women, Work, and the Family from Slavery to the Present.* New York: Basic Books, 1985.

Katz, William Loren. *Black People Who Made the Old West.* New York: Crowell, 1977.

———. *The Black West.* 1971. Rpt. Seattle, Wash.: Open Hand Publishers, 1987.

———. *Black Women of the Old West.* New York: Atheneum, 1995.

Keane, Melissa. "Cotton and Figs: The Great Depression in the Casa Grande Valley." *Journal of Arizona History* 32:3 (Autumn 1991): 267–90.

Keegan, Frank L. *Black Towns, USA.* Boston: Little Brown, 1971.

Kendrich, Gregory D. *Black Indians: A Hidden Heritage.* New York: Atheneum, 1986.

Kester, Howard. *Revolt among the Sharecroppers.* New York: Arno Press, 1969.

Koos, Earl S. *They Follow the Sun.* Jacksonville: Florida State Board of Health, 1957.

Larson, Ernie. "Migration and Its Effect on Schools." *Elementary School Journal* 41:295 (December 1940).

Lavers, George. "State School for Girls Shows Great Promise." *Casa Grande Valley Dispatch,* October 27, 1933.

Leckie, William H. *The Buffalo Soldiers: A Narrative of the Negro Cavalry in the West.* Norman: University of Oklahoma Press, 1967.

LeSeur, Geta. "Dreams Deferred: The Women of Randolph, Arizona." *Compendium* 2.5 (February 1989): 14–15.

Love, Nat. *The Life and Adventures of Nat Love, Better Known in the Cattle Country as "Deadwood Dick," by Himself.* 1907. Rpt. New York: Arno Press, 1968.

Lowe, Charlotte. "True West: Black Cowboy's Pioneer Spirit Has Built Ranch Town of Keeylocko." *Tucson Citizen,* October, 26, 1987, sec. D, pp. 1, 4.

MacTavish, Caton. "Randolph's Vision of Railway: Work of Pioneer Builder Epes Randolph, Recalled by Representative." *Tucson Citizen,* January 6, 1985.

Mann, Susan Archer. "Social Change and Sexual Inequality: The Impact of the Transition from Slavery to Sharecropping on Black Women." Working paper, Center for Research on Women, Memphis State University, 1986.

Martin, Patricia Preciado. *Songs My Mother Sang to Me: An Oral History of Mexican American Women.* Tucson: University of Arizona Press, 1992.

Melcher, Mary. "Christine E. Hankins; Geneva Hankins." Interviews, Casa Grande Historical Society, May 28, 1992.

"Migrants Show Decline in Numbers." *Coolidge Examiner,* January 2, 1948, p. 1.

Moore, Truman. *The Slaves We Rent.* New York: Random House, 1965.

Morrison, Toni. *Paradise.* New York: Alfred A. Knopf, 1998.

Otey, Frank M. *Eatonville, Florida: A Brief History of One of America's First Freedmen's Towns.* Winter Park, Fla.: Four G Publishers, 1976.

Padfield, Harlan, and William E. Martin. *Farmers, Workers, and Machines: Technological and Social Change in Farm Industries in Arizona.* Tucson: University of Arizona Press, 1965.

Painter, Nell Irwin. *Exodusters.* New York: Norton, 1976.

"Pickers Paid Millions." *Coolidge Examiner,* December 31, 1948, p. 1.

Piggott, Lyle. "Randolph: Tiny Pinal Community Dates Back to 1920s." *Tri-Valley Dispatch,* May 7, 8, 1986, pp. 1–2.

Porter, Kenneth Wiggins. *The Negro on the American Frontier.* New York: Arno Press, 1970.

"Prescott Prospectus of the Salt River Valley. . . ." *Prescott Arizona Enterprise,* April 6, 1878, sec. 2, p. 4.

"The Romance of Randolph." *Arizona Bulletin* (1925).

Richardson, Barbara. "Research on Black Women in the Early Southwest U.S." Manuscript, 1987.

Ross, Frank Alexander. *A Bibliography of Negro Migration.* New York: Columbia University Press, 1934.

Savage, W. Sherman. *Blacks in the West.* Westport, Conn.: Greenwood Press, 1976.

Schlissel, Lillian. *Women's Diaries of the Western Journey.* New York: Schocken Books, 1982.

Sloter, James Foster. *Eloy.* N.p.: ca. 1950.

Slotkin, Richard. "Gunsmoke and Mirrors: There Up on the Screen—That's Where We Invent Our Heroes." *Life Magazine,* Special Issue, "The Wild West, Yesterday and Today, April 5, 1993, pp. 60–68.

Smith, Gloria. "Black Americana in Arizona." Manuscript, Tucson, Ariz., 1977. Revised, 1992.

Stansfield, Howard. "Randolph Tin Plant Okayed for Expansion." *Casa Grande Valley Dispatch,* May 3, 1994, pp. 1–3.

Steinbeck, John. *The Grapes of Wrath.* New York: Viking, 1939.

Taylor, Paul S. *American Exodus: A Record of Human Erosion in the Thirties.* New York: Reynal and Hitchcock, 1939.

Terrell, John Upton. *Estevanico the Black.* Los Angeles: Western Lore Press, 1968.

Thompson, John, and Cary Wolinsky. "Cotton, King of Fibers." *National Geographic* 185:6 (June 1994): 60–87.

"Two Hundred from Texas." *Prospector,* June 22, 1905, sec. 1, p. 3.

United States Congress, Subcommittee on Migratory Labor. *Migrant and Seasonal Farmworker Powerlessness.* Washington, D.C.: U.S. Government Printing Office, 1970–1971.

Walker, Curtis. "The Influence of Culture and Demography on the Colored People of McNary." Master's thesis, University of Arizona, 1975.

Walton, Anthony. *Mississippi: An American Journey.* New York: Alfred A. Knopf, 1996.

Weisiger, Marsha L. *Land of Plenty: Oklahomans in the Fields of Arizona, 1933–1942.* Norman: University of Oklahoma Press, 1995.

———. "Mythic Fields of Plenty." *Journal of Arizona History* 32:3 (Autumn 1991): 241–66.

Windsor, Merrill C. "The First Negro Pupil in Casa Grande." Report, Casa Grande Historical Society, 1978.

"Witnessing the Lives of the Delta's People." *Chronicle of Higher Education,* June 2, 1995, sec. B, p. 40.

Wloszczyna, Susan. "Film Aims to Widen Range of Westerns." *USA Today,* May 12, 1993, sec. D, pp. 1–2.

Woods, Dan. "Cotton Pickers Wrote Chapter of Valley History." *Casa Grande Valley Dispatch,* January 7, 8, 1976, pp. 1–2.

Woodson, Carter G. *A Century of Negro Migration.* 1918. Rpt. Washington, D.C.: Association for the Study of Negro Life and History, 1976.

Yantis, John. "Casa Grande Looks Back at Segregation." *Arizona Republic / Phoenix Gazette,* June 10, 1992, p. 7.

Index

(Note: The cities of Casa Grande, Coolidge, Eloy, Florence, and Randolph are not indexed, as references to them occur throughout)

About the Author

Ed Brown

Associate Professor of English, Black Studies, and Women Studies at the University of Missouri–Columbia, Geta LeSeur is the author of *Ten Is the Age of Darkness: The Black Bildungsroman* (University of Missouri Press).